FUNDAMENTALS OF DIGITAL ELECTRONICS

ROSEMARY KNOX

Siemens Energy and Automation

Formerly of
Texas State Technical Institute
Waco, Texas

Breton Publishers
Boston, Massachusetts

PWS PUBLISHERS

Prindle, Weber & Schmidt • Duxbury Press • PWS Engineering • Breton Publishers •
20 Park Plaza • Boston, Massachusetts 02116

PWS Publishers is a division of Wadsworth, Inc.

Library of Congress Cataloging-in-Publication Data

Knox, Rosemary.
  Fundamentals of digital electronics.
  Includes index.
  1. Digital electronics. I. Title.
Tk7868.D5K56 1986   621.3815   86-8100
ISBN 0-534-06402-7

Printed in the United States of America
1 2 3 4 5 6 7 8 9—90 89 88 87 86

ISBN-0-534-06402-7

Sponsoring editor: *George J. Horesta*
Production supervision: *Technical Texts, Inc.*
Production editor/Copy editor: *Jean T. Peck*
Text design/Cover design: *Sylvia Dovner*
Text illustration: *Horvath & Cuthbertson*
Composition: *Crane Typesetting Service*
Cover printing: *New England Book Components, Inc.*
Text printing and binding: *The Alpine Press*

Acknowledgment: Figure 2–6, used by permission of Hewlett-Packard, Santa Clara, CA.

Preface

Fundamentals of Digital Electronics is a practical text that is intended for use in introductory courses in the essentials of basic digital electronics. It is especially appropriate for such courses in applied, technician-oriented electronics curricula. The presentation avoids detailed coverage of theory, analysis, and design and focuses instead on circuit characteristics and operation.

It is assumed that students taking this course will have had an introductory course in basic electrical concepts and dc/ac circuits as a prerequisite. While a basic course in active devices and circuits might also be a useful prerequisite (or at least a corequisite), this course would not be absolutely essential for most students to understand the material in the presentation.

Self-test exercises are integrated with the text presentation, and review exercises appear at the end of each chapter. These tests are designed to help the student find what to study and what to review again. The answers to self-test questions are included in an answer section at the back of the book.

Laboratory experiments appear at the ends of most chapters. These experiments, in addition to emphasizing digital applications, contribute to a practical, hands-on orientation that students can readily comprehend and assimilate.

The author wishes to express her thanks to her students, her reviewers, and the staffs at PWS Publishers and Technical Texts, all of whom made this publication possible.

Contents

Digital Techniques

OBJECTIVES

After studying this chapter, you will be able to:

1. Define analog and digital signals and give examples of each.
2. Describe the advantages of digital techniques.
3. Convert binary, binary-coded decimal, octal, and hexadecimal numbers to decimal.
4. Convert decimal numbers to binary, binary-coded decimal, octal, and hexadecimal.
5. Identify logic as negative or positive when two voltage levels are given.
6. Explain the difference between serial and parallel data transmission and cite the advantages of each.

INTRODUCTION

Digital circuits have a language of their own. If we were to communicate by blinking our eyes open and closed, we would, in effect, be using digital signals. The transmission would be very slow and difficult for us. Digital circuits, however, transmit, receive, and process signals at amazing speeds. This chapter will show you how.

ANALOG AND DIGITAL SIGNALS

There are two types of electronic signals, analog and digital. An *analog signal* is an ac or dc current that varies smoothly and continuously over a period of time. That is, it does not change abruptly or in steps. A few examples of analog signals are shown in Figure 1–1. The electronic circuits that produce these signals are called *analog circuits*. Analog circuits have been designed to deal with such measurable quantities as distance, temperature, direction, fuel level, and weight. They are employed in such devices as scales and tapes, thermometers, compasses, gasoline gauges, and weighing scales.

A *digital signal* is a series of pulses or voltage levels that vary in discrete (separate or noncontinuous) steps or increments. A two-level, off–on or up–down, fast-switching character is indicative of all digital signals. The electronic circuits that produce these signals are called *digital logic*, or *pulse, circuits*. Examples of digital signals are shown in Figure 1–2. Digital circuits deal with variables that are given a numerical

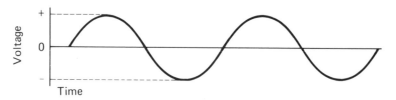

A. Sine Wave

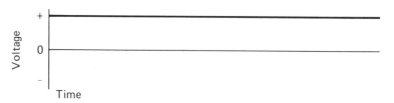

B. Positive dc Voltage

C. Varying Negative dc Voltage

FIGURE 1—1

Examples of Analog Signals

D. Random ac Voltage

value that represents one of two states. For example, auto headlights are either on or off; a camera shutter is either open or closed; and it's always one day of the week or another.

ADVANTAGES OF DIGITAL TECHNIQUES

Most computer systems are digital rather than analog because of the advantages of digital techniques. Digital techniques are used in many other systems as well—for example, communications, test equipment, process controls, and commercial electronic equipment such as TV, TV games, pinballs, vending machines, and calculators.

Digital techniques allow us to use digital logic integrated circuits. *Integrated circuits*, or ICs as they are called, are a combination of interconnecting circuit elements in a single package. Most integrated circuits are available at low cost and are extremely versatile, small, and reliable.

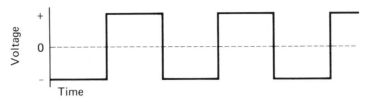

A. Square Wave

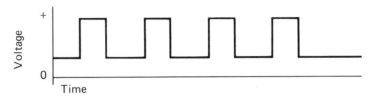

B. Positive Rectangular Wave

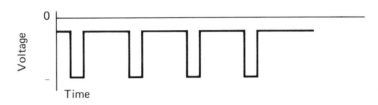

C. Negative Rectangular Wave

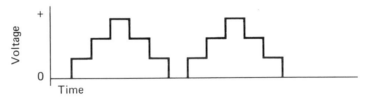

D. Digitized Analog Signal

FIGURE 1–2

Examples of Digital Signals

Some integrated circuits can replace a whole printed circuit board of analog devices and more. The cost, weight, size, and power consumption of the equipment are thereby reduced.

Digital circuits provide higher accuracy and resolution (the smallest measurable change that can be detected) than analog circuits. Thus, computer control of manufacturing and industrial processes becomes very desirable.

The complexity—that is, the number of functional components in a small unit—of circuitry often increases when digital techniques are used. Therefore, the better your knowledge and understanding of these techniques, the more quickly you will see that digital circuits are not necessarily difficult to troubleshoot or repair.

SELF-TEST EXERCISE 1–1

1. What are the two types of electronic signals?
2. Analog signals vary _____. Digital signals vary _____.
3. Identify the following as digital or analog:
   a. decimal numbers
   b. telephone dial
   c. volume control on radio
   d. gas tank gauge
4. What are the special features of integrated circuits?
5. What are the advantages of using ICs in digital equipment?
6. A constant dc voltage can be either analog or digital depending on how it is used. True or False?
7. The waveform in Figure 1–3 is:
   a. analog
   b. digital

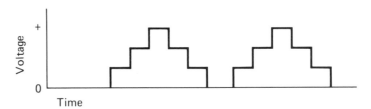

FIGURE 1–3

BINARY NUMBER SYSTEM

All digital circuits or systems work with numbers in some form or another. For example, a calculator does calculations on the numbers that are entered and gives a number as an output. It accepts the numbers, processes them, and generates number outputs. The numbers may be in binary or decimal form depending on the application.

We use the *decimal number system* in our day-to-day lives. Its ten digits, 0 through 9, which in combination represent any other number, are so familiar to us that we may find it hard to accept the *binary number system*. The binary system has two digits, or *bits*, 0 and 1. Just as we combine digits in the decimal system, these two binary digits are combined to give any other number. The base, or *radix*, of the decimal system is 10; for binary, it is 2. Thus, binary is sometimes referred to as *radix 2* and decimal as *radix 10*.

For the decimal system, the hardware or circuits would have ten individual steps to represent digits 0 through 9. For the binary system, two steps, such as on–off or up–down, are required. These steps can be easily accomplished with relay contacts and switches, which is how the first computers were built. Solid state transistor circuits are frequently used in modern computers because they use little power and are easily switched on and off.

Decimal-to-Binary Conversion

There is more than one way to convert from decimal to binary. The following conversion method may be used not only for binary but also for any other number system you may encounter. But, to avoid confusion, it is the only method presented here. You may want to study other modes of conversion when you feel comfortable with the method in this book.

To begin, note that both decimal and binary systems are weighted or *positional* number systems. That is, each digit or bit has a different weight depending on the position it occupies in a number. In the decimal system, remember, for example, that the decimal number 1524 has four 1s, two 10s, five 100s, and one 1000:

$$1524 = (1 \times 10^3) + (5 \times 10^2) + (2 \times 10^1) + (4 \times 10^0)$$
$$= (1 \times 1000) + (5 \times 100) + (2 \times 10) + (4 \times 1)$$
$$= 1000 + 500 + 20 + 4 = 1524$$

What you may not remember is that all number systems have a *weight chart*. The decimal weight chart for the number 1524 is as follows:

Row 1	10^3	10^2	10^1	10^0	Exponent value
Row 2	1000	100	10	1	Decimal value
Row 3	1	5	2	4	Decimal digit (1524)

Row 1 contains the base number (here it is 10) with exponents that start at zero. Row 2 gives the decimal values of the numbers in row 1. Row 3 is where we place the digits that represent how many 1s, 10s, 100s, 1000s, and so on, that we need for the numeral.

Conversion in the binary system works in much the same way. That is, you always begin conversion by drawing a weight chart for the base system you want to use and writing the base number in each square of the first row. Begin with base 2, the binary number system base. The chart can go on forever in both directions, but for now, eight blocks across are enough. (Later we will see why eight blocks were chosen this time.)

Now, add the exponents to each base number, starting at zero, until you run out of blocks. Always start at the right, or least significant bit location. The *least significant bit* (LSB) is always the one that carries the least weight—2^0 in this case. Your first row should look like the following:

2^7	2^6	2^5	2^4	2^3	2^2	2^1	2^0	Exponent value

Next, calculate the decimal value of these numbers and put the values in the second row of the weight chart. Use the y^x function of your calculator to do the work. For example, to calculate 2^0,

1. Enter the numeral 2 on your calculator.
2. Press the y^x key.
3. Press zero, which is the exponent for the first box in the chart (LSB).
4. Press the equals sign ($=$) key.
5. Your calculator will display the numeral 1 for this example because any number to the zero power is 1. Write this number in the chart.

Following the preceding steps, change the exponent in step 3 for all values in the chart. Your binary weight chart should then look like the following one:

2^7	2^6	2^5	2^4	2^3	2^2	2^1	2^0	Exponent value
128	64	32	16	8	4	2	1	Decimal value

■ **EXAMPLE**

Convert the decimal number 85 to a binary equivalent numeral by using the following binary weight chart:

2^7	2^6	2^5	2^4	2^3	2^2	2^1	2^0	Exponent value
128	64	32	16	8	4	2	1	Decimal value
								Binary value

Note that the chart is eight blocks wide here because the number 128 is larger than the number 85 being converted. Any number used in the chart from now on must be binary; therefore, only the digits 0 and 1 can be used.

Solution

Beginning with the 2^6 (64) block, put a 1 in the 64s place. Since a weight of 64 has now been used, subtract 64 from the 85 you started with:

$$85 - 64 = 21$$

Look at the chart again and find the next number that does not exceed the remaining weight of 21. It is 16. A number must occupy the 32s place, however, just as the zeros hold a place in the decimal number 1000. So, put a 0 in the 32s place. Now, use a weight of 16 by placing a 1 in the 16s place. A 1 in the 16s place means that a weight of 16 has been used. So, subtract 16 from the previous weight (21):

$$21 - 16 = 5$$

Since the 8s place exceeds the remaining weight, put a 0 in the 8s place to hold it and a 1 in the 4s place to use a weight of 4. Finally, since $5 - 4 = 1$, put a 0 in the 2s place and a 1 in the 1s place. The completed chart that follows shows the conversion of 85_{10} to 1010101_2:

2^7	2^6	2^5	2^4	2^3	2^2	2^1	2^0	Exponent value
128	64	32	16	8	4	2	1	Decimal value
	1	0	1	0	1	0	1	Binary value for 85_{10}

■

■ **EXAMPLE**

Convert the decimal number 67 to binary.

Solution

Again using a weight chart, we see that one 64 is needed, one 2, and one 1. We place a 1 in each of these blocks and a 0 in the blocks not needed. The binary equivalent for 67 is, therefore, 1000011, as shown below:

2^7	2^6	2^5	2^4	2^3	2^2	2^1	2^0	Exponent value
128	64	32	16	8	4	2	1	Decimal value
	1	0	0	0	0	1	1	Binary value for 67_{10}

■

Note that the binary number system is not limited to whole number conversions. To convert the decimal number 23.25 to binary, for example, the chart must be expanded to the right to include binary fractions. In order to expand to the right, start with the exponent value 2^{-1}, then 2^{-2}, and so on, as far as is needed. Then determine the decimal value of these numbers. For example, to calculate 2^{-1}, use your calculator as follows:

1. Enter the numeral 2 on your calculator.
2. Press the y^x key.
3. Enter the numeral 1, which is the exponent.
4. Press the plus/minus sign ($\pm$) key.
5. Press the equals sign ($=$) key.
6. Your calculator will display 0.5, which is 2^{-1}. Write this number in the chart.

Following the preceding steps, change the exponent in step 3 for all values to the right of the binary point. Your chart should then look like the following one:

Binary point

2^7	2^6	2^5	2^4	2^3	2^2	2^1	2^0	2^{-1}	2^{-2}	2^{-3}	Exponent value
128	64	32	16	8	4	2	1	.5	.25	.125	Decimal value
											Binary value

To continue the conversion of the decimal number 23.25, one 16, no 8s, one 4, one 2, and one 1 are needed to equal 23. The fraction 0.25

is also needed. Fill in the chart with a 0 in the 0.5 block and a 1 in the 0.25 block. The conversion for 23.25 is, therefore, 10111.01, as shown below:

Binary point

2^6	2^5	2^4	2^3	2^2	2^1	2^0	2^{-1}	2^{-2}	2^{-3}	Exponent value
64	32	16	8	4	2	1	.5	.25	.125	Decimal value
		1	0	1	1	1	0	1		Binary value for 23.25_{10}

Note that conversions to three decimal places are also possible. For example,

$$67.625_{10} = 1000011.101_2$$

$$36.875_{10} = 100100.111_2$$

$$103.125_{10} = 1100111.001_2$$

Figure 1–4 shows the conversions for the preceding numbers.

Binary point

2^6	2^5	2^4	2^3	2^2	2^1	2^0	2^{-1}	2^{-2}	2^{-3}	Exponent value
64	32	16	8	4	2	1	.5	.25	.125	Decimal value
1	0	0	0	0	1	1	1	0	1	Binary value for 67.625
	1	0	0	1	0	0	1	1	1	Binary value for 36.875
1	1	0	0	1	1	1	0	0	1	Binary value for 103.125

FIGURE 1–4

Binary Weight Chart for Decimal-to-Binary Conversion

Binary-to-Decimal Conversion

Conversion from binary to decimal is even easier than converting from decimal to binary. Place a binary number in the binary weight chart, and then add the decimal values of the blocks containing a 1 to obtain the equivalent decimal value.

■ EXAMPLE

Convert the binary number 10110 to decimal.

Solution

No 1s, one 2, one 4, no 8s, and one 16 are used. Therefore,

$$2 + 4 + 16 = 22$$

Thus, the equivalent decimal value is 22.   ■

Binary Words

Binary numbers are also called *binary words*. An 8-digit binary number is an 8-bit word, or one *byte*. That is, it has eight positions in the weight chart. It can represent a decimal number from 0 through 255, as the following three examples show:

$$0_{10} = 00000000_2$$

$$255_{10} = 11111111_2$$

$$135_{10} = 10000111_2$$

The subscript 2 indicates the number is in base 2. A subscript is seldom used in base 10 because it is assumed. Figure 1–5 shows the binary-to-decimal conversion of each example. Study the chart; follow the steps as you did for converting 85 to 1010101_2.

8	7	6	5	4	3	2	1	Position of digits
2^7	2^6	2^5	2^4	2^3	2^2	2^1	2^0	Exponent value
128	64	32	16	8	4	2	1	Decimal value
0	0	0	0	0	0	0	0	0_{10}
1	1	1	1	1	1	1	1	255_{10}
1	0	0	0	0	1	1	1	135_{10}

FIGURE 1–5

Binary Weight Chart for Binary-to-Decimal Conversion

The maximum number of states, or *discrete steps*, that can be represented with eight bits is 256. (Each number is a discrete step. Remember, zero is a number too.) The following formula will help you to quickly determine how many states are possible for a given number of bits:

$$\text{maximum number of states} = 2^N$$

where N is the number of bits or digits in a number. For example, for eight bits or an 8-bit word,

$$2^N = 2^8 = 256 \text{ states}$$

Since zero is one of these states, the maximum decimal number we can represent is $256 - 1$, or 255. To show that $2^N - 1 =$ maximum decimal number for an N-bit word, let us look at another example.

■ EXAMPLE

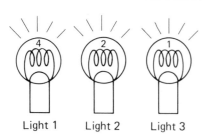

Light 1   Light 2   Light 3

FIGURE 1–6

3-Bit Binary Lights

Let the three light bulbs in Figure 1–6 represent the three bits of a 3-bit word. How many different combinations of "on" and "off" can be created?

Solution

Figure 1–7 shows all the possibilities in tabular form. You can see that there are only eight different combinations, which agrees with the formula given earlier:

$$2^N = 2^3 = 8 \text{ states}$$

Again, zero is a state so that when you assign each of these states a number, starting with all lights off as 0 and proceeding to all lights on as 7, you find that the maximum decimal number these lights can represent is

$$2^N - 1 = 8 - 1 = 7$$

Light 1	Light 2	Light 3	Combinations or discrete states	Decimal equivalent
Off	Off	Off	State 1	0
Off	Off	On	State 2	1
Off	On	Off	State 3	2
Off	On	On	State 4	3
On	Off	Off	State 5	4
On	Off	On	State 6	5
On	On	Off	State 7	6
On	On	On	State 8	7

FIGURE 1–7

3-Bit Binary Truth Table

The minimum number of bits required to represent a decimal number as a binary number can also be calculated.

■ EXAMPLE

How many bits are needed to represent the decimal number 33 in binary?

Solution

The number 33 is converted to binary as follows:

2^6	2^5	2^4	2^3	2^2	2^1	2^0	Exponent value
64	32	16	8	4	2	1	Decimal value
0	1	0	0	0	0	1	Binary value for 33_{10}

Count the "binary places" in the same way that you count decimal places. There are six, so it takes six places or bits to represent 33 in binary. Note that it also takes a minimum of six bits to represent the decimal numbers 32 to 63 (100000 to 111111). ■

The following formula can also be used to determine the minimum number of bits needed to represent a decimal number in binary. This formula is useful only up to nine bits, however.

$$\text{minimum number of bits} = 3.32 \log_{10}(N)$$

where N is the decimal value.

■ **EXAMPLE**

How many bits are needed to represent the decimal number 144 in binary?

Algebraic Solution
Solve for x in the formula:

$$3.32 \log_{10}(144) = x$$
$$3.32(2.158) = x$$
$$7.16 = x$$

Since you cannot use less than a whole bit, round off to the next whole value, or 8. It takes eight bits to represent 144 in binary.

Calculator Solution
To use your calculator to solve for x,

1. Enter the number 144.
2. Press the log function key.
3. Press the multiplication sign ($\times$) key.
4. Enter the number 3.32.
5. Press the equals sign ($=$) key. ■

It is beyond the scope of this book to explain how this formula came about. It is a handy tool, however, and you should memorize it.

Conversion from a decimal number to a binary number is called *coding*. Since columns of binary 0s and 1s, or *pure binary code*, can be very cumbersome to work with, other binary codes have been developed for ease of use. These include, for example, the Binary-Coded Decimal (BCD), Excess 3, Gray, and ASCII codes. It is important to understand BCD code since it is the most commonly used coding for computer circuits.

BINARY-CODED DECIMAL SYSTEM

The *Binary-Coded Decimal System* is a compromise between the decimal and the binary systems. The decimal system is easy for us to use because we are so familiar with it, and the binary system is easy for digital circuits to work with because binary digits represent two distinct voltage levels. The BCD system was developed because it is easier to understand and convert than pure binary code.

Decimal-to-BCD Conversion

The BCD system represents decimal digits 0 through 9 with a 4-bit binary code in the 8–4–2–1 position weighting system. You can substitute the individual digits by their binary code to represent any decimal number in BCD. Figure 1–8 illustrates the following example:

$$156_{10} = 0001 \quad 0101 \quad 0110_{BCD}$$

Note that a gap is left between the digits to avoid confusion with the format of pure binary code. Each group of 8–4–2–1 represents just one decimal position. As with the weight charts for pure binary, BCD weight charts can also go on forever.

FIGURE 1–8

BCD Weight Chart

Decimal places	100s				10s				1s			
8421	8	4	2	1	8	4	2	1	8	4	2	1
BCD value for 156_{10}	0	0	0	1	0	1	0	1	0	1	1	0
Decimal value	1				5				6			

Advantages and Disadvantages of BCD

One advantage of BCD code is the ease with which numbers can be changed from this system to others. Although the BCD system is very helpful in simplifying man-to-machine interface, it is less efficient compared to the pure binary code. The number of bits required to represent a decimal number in BCD is more than in pure binary code. For example,

$$96_{10} = 1100000_{binary} = 1001 \quad 0110_{BCD}$$
$$126_{10} = 1111110_{binary} = 0001 \quad 0010 \quad 0110_{BCD}$$

Furthermore, a certain amount of extra digital circuitry is required for every extra bit. So, BCD has its disadvantages: higher costs, bigger systems, and greater power consumption.

OCTAL NUMBER SYSTEM

The *octal number system*, base 8 or radix 8, is another important number system. It is related to the binary system in that $2^3 = 8$. It was used for entering data in early computers.

In the octal system, eight individual steps are used to represent the digits 0 through 7. These eight digits can be combined to represent any number. The subscript 8 is used to identify a number as octal.

Decimal-to-Octal Conversion

Conversion from decimal to octal is done in the same way as binary conversions are done except that the weight digits include 0 through 7. This conversion method is a little more confusing, but the following example should make its logic easier to perceive.

■ **EXAMPLE**

Convert the decimal number 85 to base 8 using an octal weight chart.

Solution
1. Since there is one 64 in the number 85, enter a 1 in the 64s place. Subtract 64 from 85; a weight of 21 remains.
2. Since there are two 8s in 21 [$21 - (2 \times 8) = 5$], enter a 2 in the 8s place. Subtract 16 from 21; a weight of 5 remains.
3. Since there are five 1s in 5 [$5 - (5 \times 1) = 0$], enter a 5 in the 1s place.

The conversion of 85_{10} is, therefore, 125_8, as shown below:

3	2	1	0	Position of digits
8^3	8^2	8^1	8^0	Exponent value
512	64	8	1	Decimal value
Not needed	1	2	5	Octal value for 85_{10}

Check
Multiply all the octal digits in the weight chart by the decimal value they represent. Then, add all these values together:

$$(1 \times 64) + (2 \times 8) + (5 \times 1) = 85$$

$$64 + 16 + 5 = 85$$

$$85 = 85$$

■

Figure 1–9 shows the decimal-to-octal conversions for the following numbers:

$$3346_{10} = 6422_8$$

$$596_{10} = 1124_8$$

$$67_{10} = 103_8$$

Study the chart; follow the steps as you did for converting 85_{10} to 125_8.

8^3	8^2	8^1	8^0	Exponent value
512	64	8	1	Decimal value
6	4	2	2	3346_{10}
1	1	2	4	596_{10}
0	1	0	3	67_{10}

FIGURE 1–9

Octal Weight Chart

HEXADECIMAL NUMBER SYSTEM

The *hexadecimal number system* is often used as the input code or program code for microcomputers like the Apple, IBM PC, and Commodore. It is also related to the binary system in that $2^4 = 16$.

In the hexadecimal system, the base or radix is 16. Sixteen weight digits are thus required. The digits 0 through 9 are used; but to represent the decimal equivalent of 10, 11, 12, 13, 14, and 15, a single digit value is necessary. The letters A through F are used for simplicity. They are commonly found on printers, typewriters, and keyboards and eliminate the need for special symbols. The radix 16 symbols and their decimal equivalents are as follows:

Decimal equivalent	0	1	2	3	4	5	6	7	8	9	10	11	12	13	14	15
Radix 16 symbol	0	1	2	3	4	5	6	7	8	9	A	B	C	D	E	F

A radix or base 16 weight chart is shown in Figure 1–10. The subscript 16 is used to identify a number as hexadecimal.

4	3	2	1	Position of digits
16^3	16^2	16^1	16^0	Exponent value
4096	256	16	1	Decimal value
1	3	B	C	Hexadecimal value for 5052_{10}
4096	768	176	12	Decimal equivalent

FIGURE 1–10
Hexadecimal Weight Chart

Decimal-to-Hexadecimal Conversion

Conversion from decimal to hexadecimal is done in the same way as binary conversions are done except that there are sixteen weight digits, 0 through F. The following example illustrates the conversion.

■ **EXAMPLE**

Convert the decimal number 5052 to base 16. See the hexadecimal weight chart in Figure 1–10.

Solution

1. Since there is one 4096 in the number 5052, enter a 1 in the 4096s place. Then subtract:

$$5052 - 4096 = 956$$

2. Since there are three 256s in 956, enter a 3 in the 256s place. Then subtract:

$$956 - (3 \times 256) = 188$$

3. Since there are eleven 16s in 188, enter an 11 in the 16s place. You cannot enter two digits, of course; so use B to represent the 11. And subtract:

$$188 - (11 \times 16) = 12$$

4. Since there are twelve 1s in 12, enter a 12—that is, a C—in the 1s place.

The conversion is complete: $5052_{10} = 13BC_{16}$.

Check

$$(1 \times 4096) + (3 \times 256) + (B \times 16) + (C \times 1) = 5052$$
$$(1 \times 4096) + (3 \times 256) + (11 \times 16) + (12 \times 1) = 5052$$
$$4096 + 768 + 176 + 12 = 5052$$
$$5052 = 5052 \quad ■$$

As you work with the different number systems, you will begin to see the relationship of one system to another. Conversion from one number system to any other will be accomplished with ease once these relationships are realized.

SELF-TEST EXERCISE 1–2

1. What is the base or radix of binary systems?
2. How many discrete steps are in binary numbers?
3. Convert the following decimal numbers to binary:
   a. 10
   b. 35
   c. 100
   d. 107.75
4. Convert the following binary numbers to decimal:
   a. 11011_2
   b. 110111_2
   c. 1111111_2
   d. 100001.011_2
5. What is the maximum decimal number that can be represented by six binary digits?
6. How many states are there for six binary digits?
7. How many bits are required to represent 278 in binary?
8. What is BCD?
9. Represent the following decimal numbers in BCD:
   a. 126
   b. 723
   c. 15
   d. 9
10. Convert the following base 8 numbers to decimal:
    a. 127_8
    b. 523_8
    c. 1064_8
11. Convert the following decimal numbers to octal:
    a. 1093
    b. 456
    c. 93
12. Convert the following hexadecimal numbers to decimal:
    a. $A23_{16}$
    b. FF_{16}
    c. BCA_{16}
13. Convert the following decimal numbers to hexadecimal:
    a. 432
    b. 5564
    c. 98
14. The binary system uses digits 0 through 9. True or False?
15. The number 1111 in binary is equal to 15 in decimal. True or False?
16. The largest decimal number represented by an 8-bit binary number is 255. True or False?
17. The number 100_{10} can be represented by seven bits in binary. True or False?
18. The number 100_{10} is equivalent to 1100101 in binary. True or False?
19. BCD uses the digits 0 through 9. True or False?
20. The number 10_{10} is 0001 in BCD. True or False?
21. The number 10_{10} is 1010 in BCD. True or False?

HARDWARE USED TO REPRESENT DATA

Electronic components and circuits that are used to represent and manipulate binary numbers in a system are called *hardware*. The components used should be capable of assuming two distinct voltage levels, represented by binary 0 and binary 1.

Electromechanical devices such as relays and switches are used to represent static (unchanging) and slow-speed binary conditions. An open switch contact can represent a binary 0, and a closed switch contact can represent a binary 1. The states can be reversed so that an open switch is a binary 1 and a closed switch is a binary 0.

Electronic devices such as vacuum tubes and transistors are used to represent the binary states where high speed is required. A tube or transistor when it is either conducting or cut off can be used to represent the 0 or 1 state. Vacuum tubes, however, have been replaced by transistors because of their reliability, low cost, and low power consumption. Transistors offer a high resistance when they are not conducting, equivalent to an open switch, and a very low resistance when they conduct heavily, similar to a closed switch. Transistors are still used in digital equipment and computers. (Their use will be more apparent later in this text.) The most common type of transistor used is the *bipolar* or *metal-oxide semiconductor field effect transistor* (MOSFET). This text will deal primarily with bipolar transistor circuits and transistor–transistor logic (TTL). Their durability and minimum power supply requirements are reasons enough for their use.

LOGIC LEVELS

We know that a binary word is a string of 1s and 0s and that hardware is the electronic circuitry that produces these 1s and 0s. The voltages or levels of binary words determine whether the electronic hardware recognizes the signals as real information. A discussion of logic levels follows.

There are two basic types of logic levels, positive logic and negative logic. When the most positive voltage level is defined as the binary 1 state, it is called *positive*, or *true*, *logic*. When the most negative voltage or the least positive voltage level is defined as the binary 1 state, it is called *negative*, or *false*, *logic*. See Figure 1–11. *Positive logic is assumed in this text unless specified otherwise.*

Positive	Negative
+5 V binary 1 0 V binary 0	0 V binary 1 +5 V binary 0
+15 V binary 1 -2 V binary 0	+15 V binary 0 -1 V binary 1

FIGURE 1–11

Positive and Negative Logic

The basic element that represents a binary bit of data is a *switch*: mechanical, electromechanical, electronic, or magnetic. We are not concerned with whether the switch is on or off in an actual situation. Instead, we must remember that the bits are represented by voltage levels. A binary 0 will be low voltage, negative voltage, or ground in positive logic. A binary 1 will be 3 V (or more) more positive than the binary 0 depending on the circuits and power supplies used. The experiments in this book use TTL circuits in which a binary 1 is +2.4 V to +5.0 V and a binary 0 is 0 V to +0.8 V.

DIGITAL DATA TRANSMISSION

Digital data is transmitted or processed in two basic ways, serial and parallel. In *serial mode*, a data word is sent or processed one bit at a time. In *parallel mode*, all the bits of a word are transmitted or processed at the same time.

Serial Transmission

The primary advantage of serial transmission is that it needs only one line to transmit data. Since only one bit is available for processing at a time, only one set of digital circuitry is needed to process the data. Thus, serial transmission is economical and simple, but the speed is poor. The system is frequently used, however, because of its economy and simplicity and in spite of its slow speed. For example, data sent over telephone lines is transmitted serially, least significant bit (LSB) first, most significant bit (MSB) last. See Figure 1–12, in which 1001 is sent in serial mode.

FIGURE 1–12

Serial Transmission

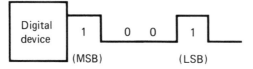

Parallel Transmission

All bits of a word are transmitted at the same time in a parallel data system. Therefore, the number of channels required depends on the number of bits in a word. A set of digital circuits is required for processing each bit; so the system becomes expensive and complex. The advantage is speed. The parallel system is used for rapid processing. The 4-bit word 1001 is transmitted in a parallel system as shown in Figure 1–13.

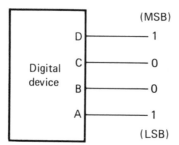

FIGURE 1–13

Parallel Transmission

ASCII CODE

One of the most common communication codes used to interface hardware between computers and keyboards, printers, video displays, and other peripheral equipment is the *American Standard Code for Information Interchange* (ASCII). This code is classified as an alphanumeric code because it contains both letters and numbers. It also contains codes for control signals to peripheral devices.

The ASCII code uses seven binary bits. The code is transmitted one bit at a time in serial or one word at a time in parallel mode. A partial listing of the ASCII code is shown in Table 1–1. This table also contains the octal and hexadecimal (hex) equivalent for each code.

TABLE 1–1

ASCII Code

Character	7-Bit ASCII Code	Octal	Hex
A	100 0001	101	41
B	100 0010	102	42
C	100 0011	103	43
D	100 0100	104	44
E	100 0101	105	45
F	100 0110	106	46
G	100 0111	107	47
H	100 1000	110	48
I	100 1001	111	49
J	100 1010	112	4A
K	100 1011	113	4B
L	100 1100	114	4C
M	100 1101	115	4D
N	100 1110	116	4E
O	100 1111	117	4F
P	101 0000	120	50
Q	101 0001	121	51
R	101 0010	122	52
S	101 0011	123	53
T	101 0100	124	54
U	101 0101	125	55
V	101 0110	126	56
W	101 0111	127	57
X	101 1000	130	58
Y	101 1001	131	59
Z	101 1010	132	5A
0	011 0000	060	30
1	011 0001	061	31
2	011 0010	062	32
3	011 0011	063	33
4	011 0100	064	34
5	011 0101	065	35
6	011 0110	066	36
7	011 0111	067	37
8	011 1000	070	38
9	011 1001	071	39
BLANK	010 0000	040	20
(	010 1000	050	28
+	010 1011	053	2B
$	010 0100	044	24
*	010 1010	052	2A
)	010 1001	051	29
−	010 1101	055	2D
/	010 1111	057	2F
,	010 1100	054	2C
=	011 1101	075	3D

DIGITAL CIRCUITS

Basically, the transistor is used in digital circuits with other components to manipulate or to process binary input data. There are two basic types of logic circuits, *decision-making* and *memory*. These circuits accept data and, based on the information, either generate output signals with the decision-making ability or temporarily store data for later use with the memory ability.

Decision-making circuits are called *gates*. They have one or more inputs and a single output. They accept binary inputs, and they output binary data. These gates can be combined in a variety of ways to make complicated decisions.

Memory circuits are called *flip-flops*. They are used to store binary data. Flip-flops are combined for a class of circuits known as *sequential circuits*. These circuits store, count, manipulate, and shift binary data.

Digital circuits in electronic equipment use both gates and flip-flops to form combinational and sequential circuits. The basic components used are transistors and diodes combined with resistors, capacitors, and other devices to produce various logic functions. Today, most digital logic is in integrated circuit (IC) form. ICs are entire logic circuits built in a single chip. Details about decision-making and memory elements will be presented in later chapters.

SELF-TEST EXERCISE 1–3

1. What is hardware?
2. How does the transistor represent the two states in binary?
3. What are the most common types of transistors used to represent binary data?
4. Define the term *positive logic*.
5. Define the term *negative logic*.
6. Identify the following voltage levels as positive or negative logic signals:
   a. -5 V = 1, $+1$ V = 0
   b. 0 V = 1, -5 V = 0
   c. $+0.3$ V = 0, $+10$ V = 1
   d. $+3$ V = 0, 0 V = 1
7. List the two different types of digital data transmission.
8. Serial data is: (Choose one or more answers)
   a. economical
   b. slow
   c. expensive
   d. fast
   e. none of the above
9. A flip-flop is: (Choose one or more answers)
   a. a memory circuit
   b. part of a sequential circuit
   c. a decision-making gate
   d. none of the above
10. Decision-making circuits are:
    a. flip-flops
    b. counters
    c. gates
    d. all of the above
11. Sketch the binary waveform of a serial data word for the number 20 where the LSB is transmitted first. Assume *negative* logic assignments of the binary levels 0 V and $+5$ V.

12. Hardware is the code used in binary systems. True or False?
13. The digits 0 and 1 are the same voltage levels for all digital circuits. True or False?
14. Positive logic means that the most positive voltage level is binary 1. True or False?
15. +5 V = 1, +0 V = 0 is positive logic. True or False?
16. +5 V = 0, +1 V = 1 is negative logic. True or False?
17. −5 V = 1, +1 V = 0 is negative logic. True or False?
18. Serial transmission transmits data at high speed. True or False?
19. In serial data, the binary bits of a word are sent one bit at a time. True or False?
20. To process parallel data, complex circuitry is required. True or False?
21. Parallel data transmission is used for high-speed processing. True or False?
22. Gates are decision-making elements. True or False?
23. Memory can be flip-flops. True or False?

SUMMARY

An analog signal is one that varies smoothly over a period of time. A digital signal is one that changes in abrupt or discrete steps. Binary digital signals are either high or low, 1 or 0. The voltage levels of each depend on the type of circuit being used.

Digital techniques offer many advantages. Many integrated circuits are available at low cost, and they are versatile, reliable, and small.

Conversion of decimal numbers to binary, BCD, octal, and hexadecimal is easily done by constructing a simple weight chart. Conversion of binary, BCD, octal, and hexadecimal numbers to decimal is even easier when a weight chart is used.

Positive logic exists when the most positive voltage level is defined as binary 1. Negative logic exists when the most negative voltage level is defined as binary 1.

Serial data transmission occurs when binary digits are transmitted one at a time over a single line. Parallel data transmission occurs when all bits of a binary word are transmitted at once, with each bit having its own line.

CHAPTER 1
REVIEW EXERCISES

1. What is a constant dc voltage?

2. The waveform in Figure 1–14 is:

    a. analog
    b. digital

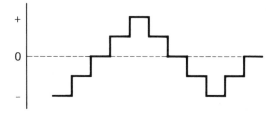

FIGURE 1–14

3. Identify each item below as analog or digital:

    a. auto headlights
    b. weather vane
    c. gasoline gauge
    d. camera shutter

4. What factor has influenced the growth and increased use of digital techniques more than any other?

5. Which of the following is not an advantage of digital over analog techniques?

    a. greater accuracy
    b. simplicity
    c. greater dynamic range
    d. better stability

6. Convert the following binary numbers to decimal:

    a. 1001011_2
    b. 1110110010_2

7. Convert the following decimal numbers to binary:

    a. 1000
    b. 95

8. Convert the following 8–4–2–1 BCD numbers to decimal:

    a. $1000 \quad 0110 \quad 0010 \quad 0101_{BCD}$
    b. $0001 \quad 1001 \quad 0111 \quad 0100_{BCD}$

9. Convert the following decimal numbers to 8–4–2–1 BCD:

    a. 893
    b. 2486

10. Convert the following decimal numbers to radix 8 (base 8):

    a. 594
    b. 63
    c. 4097

11. Convert the following base 8 numbers to decimal:

    a. 1070_8
    b. 111_8

12. Convert the following decimal numbers to hexadecimal:

    a. 396
    b. 500

13. Convert the following hexadecimal numbers to decimal:

    a. $A17_{16}$
    b. $FF4_{16}$

14. The highest decimal number that can be represented with ten bits is _____.

15. It takes _____ binary bits to represent the decimal number 121.

16. Which type of logic do digital signal levels of binary $0 = -0.7$ V, binary $1 = -1.7$ V represent?

   a. positive
   b. negative

17. Five wires, A through E, carry the voltage levels that represent a binary number. These levels are A = 0 V, B = 0 V, C = +5 V, D = 0 V, and E = +5 V. Assuming that E is the MSB and that positive logic is used, what decimal number does this signal represent?

18. What is the primary advantage of serial data transmission?

19. Name three electronic devices that use digital techniques.

20. The most common way of electronically representing binary data today is with a _____.

21. Sketch the binary waveform of a serial data word for the decimal number 18 where the LSB is transmitted first. Assume negative logic assignments of the binary levels 0 V and +10 V.

Basic Logic Gates

After studying this chapter, you will be able to:

1. Identify the two basic types of logic gate circuits used in digital equipment.
2. Identify the logic symbols for AND, OR, inverter (NOT), NAND, and NOR gates.
3. Draw a truth table for each of the basic logic gates.
4. Write the Boolean equation for each of the basic logic gates.
5. Draw a timing diagram for the operation of each basic gate.
6. Breadboard a simple gate circuit and draw a truth table of its operation.
7. Define the term *refresh*.
8. Give the voltage levels, including the disallowed levels, required for TTL positive logic.
9. Draw the logic symbol and truth table for an Exclusive OR gate.
10. Identify a logic function given its Boolean equation.
11. Draw a truth table given the inputs of a logic gate.
12. Define the term *fan-out*.
13. Give the fan-out for CMOS and basic TTL ICs.
14. Describe SSI, MSI, and LSI chips.
15. Describe propagation delay.

INTRODUCTION

As the world of electronics expanded and the power and versatility of digital circuits were discovered, a way had to be found to reduce the size of the circuits. Integrated circuits (ICs) were the answer. But the cost involved in designing and manufacturing a single circuit did not permit a separate IC to be made for every application. Therefore, basic logic gates were made; more complex circuits could be built by combining these gates. This chapter examines the operation of basic gates.

LOGIC GATE CIRCUITS

The two basic types of logic circuits are combinational logic and sequential logic circuits. *Combinational logic circuits* make decisions based on binary

24

input and type of logic circuit. *Sequential logic circuits* are memory or timing circuits that can remember a one or a zero at the output after the binary input signal is removed.

Gates

The basic elements of combinational logic circuits are AND, OR, and inverter (NOT) *gates*. All combinational circuits are combinations of these gates—hence, the term *combinational*. The gates in combinational circuits are combined to perform a specific decision-making operation. Many gates can be combined to make a circuit that performs complex decisions involving decoding, multiplexing, and arithmetic operations. Combinational circuits have one or more inputs and outputs.

AND, OR, and inverter gates can also be combined to provide memory. A memory circuit involves sequential logic. The outputs of sequential logic circuits are binary in nature and are used to control or operate other circuits. To distinguish input signals from one another, they are labeled by a letter or by the function that is performed—for example, reset, R, clear, preset, S, enable, and E.

Inverter Operation

The inverter gate is often called a NOT gate. The inverter is the easiest logic gate to remember; it always gives an output that is the opposite, or the *complement*, of the input. Zero and one are complements of each other. A zero input to the gate produces a one output. A one input produces a zero output.

Truth Tables

The operation of a logic gate is summarized in the form of a *truth table*, which is a table showing all possible inputs and outputs. If any logic gate or circuit does not operate exactly as shown in its truth table, then there is a fault in the circuit.

Figure 2–1 shows the logic symbols and truth table used in schematic drawings to indicate the inverter operation. The triangle in the logic symbol represents the gate. The small circle in front of or behind the triangle indicates the NOT or inverter operation. The most common symbol shows the small circle in front of the triangle. The inverter symbol should not be confused with other similar schematic symbols such as those shown in Figure 2–2.

Boolean Equations

Equations are often written to express logic functions. These equations are called *Boolean equations*, named after George Boole, the mathematician and logician who introduced a type of algebra in 1847 that manipulated logic variables.

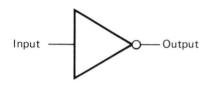

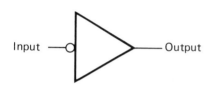

Input	Output
0	1
1	0

FIGURE 2–1

Inverter Logic Symbols and Truth Table

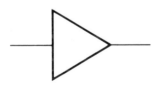

Noninverting Buffer

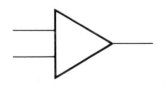

Operational Amplifier

FIGURE 2–2

Schematic Symbols Similar to Inverter Symbol

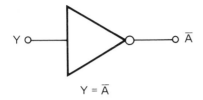

FIGURE 2–3

Inverter Symbol and Its Boolean Equation

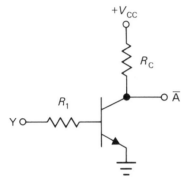

FIGURE 2–4

Analog Inverter Circuit

The Boolean equation for the inverter operation is $Y = \overline{A}$. The bar over the A means NOT, and the equation is read as "Y equals NOT A" or "Y equals the complement of A." Figure 2–3 shows the inverter symbol with its Boolean notation, indicating the output is the complement of the input.

The circuit shown in Figure 2–4 is a simple analog inverter. This circuit operates with zero as ground and some positive voltage (equivalent to supply voltage $+V_{CC}$) as one. When the input Y is at ground level (emitter–base junction is not forward biased), the transistor does not conduct; so the output is $+V_{CC}$ seen through the resistor R_C. When $+V_{CC}$ is applied to input Y (base–emitter junction is forward biased), the transistor will conduct, giving a ground level output seen through the conducting transistor. The actual voltage will be equal to the saturation voltage of the collector–emitter junction, $V_{CE(sat)}$.

Timing Diagrams

The inverter circuit in Figure 2–4 performs the inversion by giving a zero output for a one at the input and a one output for a zero at the input. Figure 2–5 shows the input and output signals plotted as voltage-versus-time waveforms. Because Figure 2–5 shows what is happening for a period of time (the x axis), it is referred to as a *timing diagram*. (We will become familiar with the timing diagrams of each basic gate later in this chapter.)

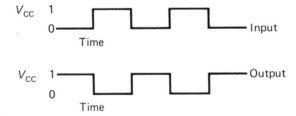

FIGURE 2–5

Inverter Timing Diagram

TROUBLESHOOTING PROCEDURES AND EQUIPMENT

Logic systems can be tested by two methods, static and dynamic. Static testing is the simpler method and should be tried first.

Performing a *static test* on an IC involves comparing the output or outputs to the truth table as the appropriate inputs are applied. Output levels can be measured with an electronic voltmeter, digital multimeter, oscilloscope, or logic probe.

A *logic probe* is designed for fast verification of digital IC inputs and outputs. It is a small, easy-to-carry, easy-to-read, and relatively inexpensive device. Several kinds are available. Figure 2–6 shows a Hewlett-Packard logic probe.

Generally, a probe derives its power from the system under test. The probe is connected between ground and the $+5$ V regulated supply of TTL ICs. Remember, a simple probe can require 5 mA to 15 mA of current to operate, which to the IC is the same as if ten TTL loads were connected. Ten loads is the maximum that a standard TTL can handle and is more than the low-power chips can properly drive. Keep this in

FIGURE 2–6

Hewlett-Packard Logic Probe

mind when you are checking circuits with a probe. The probe could cause a faulty output or even damage the chip. Therefore, buy the best probe you can afford; the better probes usually have higher input impedance, which reduces or eliminates any problem.

The condition of the signal is often displayed by three lamps located near the tip of the probe. When the probe tip is touched to a test point, the "high" lamp will indicate a one state, the "low" lamp will indicate a zero state, and the flashing lamp at the center will indicate changing or pulsing states. The specific voltages are +2.4 V to +5 V for high and 0 V to +0.8 V for low. Figure 2–7 correlates the signals on an oscilloscope to those on a probe.

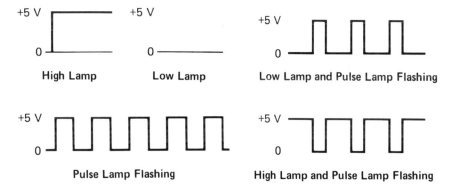

FIGURE 2–7

Logic Probe Signals

A probe that can detect a ground fault is particularly useful since this problem often occurs in a circuit. Ground fault detection means that the probe senses the difference between no input and a low or ground.

A logic clip, also shown in Figure 2–6, is a device that clips onto an IC to display the state of all pins at the same time. Like logic probes, clips too are often helpful in troubleshooting circuits.

Most of the problems you will encounter will be detectable by the static method. However, if the problem involves timing, noise, heat, or poor circuit design, the dynamic method will probably be necessary. *Dynamic testing* is performed while the circuit is in normal operation. An oscilloscope is used to look at the input and output pulses. The scope must have a triggered sweep and operate at 10 MHz or better for most circuits. A dual-trace scope is essential for comparing signals. (Other digital test equipment and troubleshooting procedures will be discussed in later chapters.)

 STOP   Do Experiment 2–1

SELF-TEST EXERCISE 2–1

1. Name the three basic decision-making elements (gates) of combinational logic circuits.
2. Digital gates output (decimal, binary) data.
3. _____ remember as long as power is applied to the circuit.
4. The NOT gate is also known as the _____ gate.
5. Many _____ circuits are combined to make complex decisions.
6. When an inverter gate input is A, the output is _____.
7. The operation of all gates can be summarized by a _____.
8. Both of the logic symbols in Figure 2–8 are symbols for the inverter gate. True or False?

FIGURE 2–8

9. The logic symbol in Figure 2–9 is the most common symbol for the _____ operation.

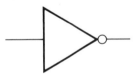

FIGURE 2–9

10. Both of the logic symbols in Figure 2–10 are symbols for the same gate. True or False?

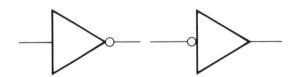

FIGURE 2–10

11. Logic systems can be tested by two methods, _____ and _____.

12. Some logic probes draw (5 mA to 15 mA, no current) from the circuit under test.

13. Buy the best test equipment you can afford. True or False?

14. (Static, Dynamic) testing is performed while the circuit is in normal operation.

15. An oscilloscope is required for all static tests. True or False?

TYPES OF LOGIC GATES

The two basic noninverting logic gates are the AND gate and the OR gate. These gates have two or more inputs and a single output. They generate a binary output depending on their input states and their function.

Two other logic gates are most often found in digital circuits, the NAND gate and the NOR gate. They are nothing more than a combination of an AND or an OR gate with an inverter (NOT–AND and NOT–OR). A discussion of the four gates follows.

The AND Gate

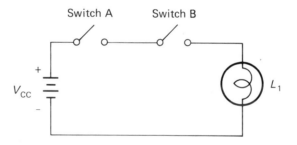

FIGURE 2–11

AND Gate Circuit

The circuit shown in Figure 2–11 will help to explain the AND gate concept. Switches A *and* B have to be closed to get light L_1 on. If we assume that

> switch closed = 1
>
> switch open = 0
>
> light on = 1
>
> light off = 0

then a truth table can be developed that summarizes all the positions of switches A and B. See Figure 2–12. Note that output $L_1 = 1$ only when A and B = 1.

Remember from Chapter 1 that the total combination of inputs is equal to 2^N, where N is the number of inputs. In this circuit, $N = 2$; so $2^2 = 4$. If there are four inputs, the number of input combinations is equal to 2^4, or 16.

A	B	L_1
0	0	0
Open	Open	Off
1	0	0
Closed	Open	Off
0	1	0
Open	Closed	Off
1	1	1
Closed	Closed	On

FIGURE 2–12

AND Gate Truth Table

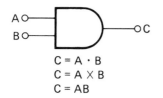

FIGURE 2–13

AND Gate Logic Symbol and Boolean Equations

The logic symbol used to represent an AND gate is shown in Figure 2–13. The Boolean equation for the AND function is $C = A \cdot B$. This Boolean equation expresses C in terms of the inputs A and B. We say that "C equals A AND B." The AND function is expressed in other ways, as shown in Figure 2–13. In this text, however, we will use the multiplication dot ($\cdot$) symbol.

More than two inputs can be used in an AND gate. For example, a three-input AND gate is expressed by the logic symbol and truth table shown in Figure 2–14. But, no matter how many inputs there are, the operation of any AND gate is the same: All inputs must be in the one state for the output to be in the one state. So, for example, all six inputs must be in the one state if an AND gate has six inputs, and the number of input combinations will equal 64.

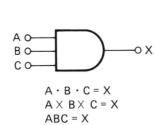

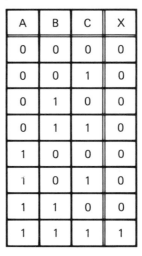

A	B	C	X
0	0	0	0
0	0	1	0
0	1	0	0
0	1	1	0
1	0	0	0
1	0	1	0
1	1	0	0
1	1	1	1

FIGURE 2–14

Three-Input AND Gate

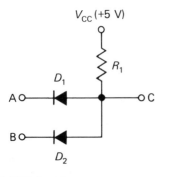

FIGURE 2–15

AND Gate Circuit Using Diodes and a Resistor

The discrete circuit in Figure 2–15 is an AND gate circuit that is formed by two diodes and a resistor and that uses binary signals. Before we examine this circuit, however, let us clarify some confusing terms. You have already learned that a one state can vary depending on the supply voltage V_{CC}. For TTL circuits, $V_{CC} = +5.0$ V; a one state is about $+2.4$ V to $+5.0$ V, and a zero state is 0 V or ground to about $+0.8$ V. Anything between these levels, $+0.8$ V to $+2.4$ V, is called *undetermined*, *ambiguous*, *indeterminate*, *disallowed*, or any other word an engineer can think of to describe the logic circuit as confused, unstable, or undependable. A one state is also expressed as *high*, H, *hi*, 1, *true*, or *yes*. A zero state is also expressed as *low*, L, *lo*, 0, *false*, or *no*.

Now, back to Figure 2–15. If the A and B inputs are high ($+V_{CC}$), no current will flow through D_1 or D_2 because they are reverse biased. Therefore, no voltage will be dropped across resistor R_1; point C will remain high ($+5$ V). If one or both A and B are grounded, a current path will be supplied through R_1; point C will be forced low. It will not be at ground level because it reflects a diode voltage drop. It is a logic low, however (less than $+0.8$ V). The timing diagrams in Figure 2–16 show the output of an AND gate with two different pulse trains as input. Study the waveforms carefully for each change.

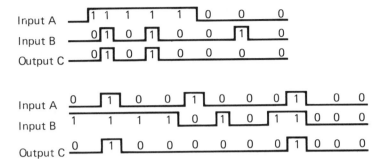

FIGURE 2–16

AND Gate Timing Diagrams

Analyzing the AND gate circuit in Figure 2–17, we see that if all or any of the inputs are low, the output will be low because the conducting transistor or transistors Q_1, Q_2, and Q_3 allow voltage to be dropped across R_1. (Note: Diode drop less than $+0.8$ V will be seen.) When all outputs go high ($+V_{CC}$), no transistors are conducting, and, thus, no current can flow. The output remains at $+5$ V potential.

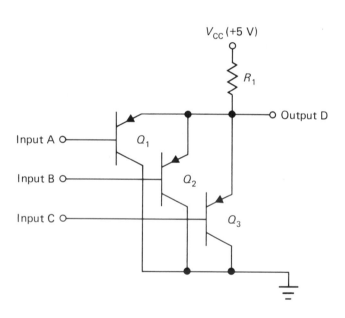

FIGURE 2–17

AND Gate Analog Circuit
Using Transistors

**SELF-TEST
EXERCISE 2–2**

1. The AND gate is a (decision-making, memory) element.
2. The circuit in Figure 2–18 (is, is not) an AND gate.

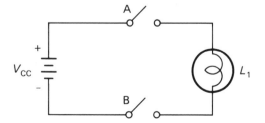

FIGURE 2–18

3. The output will be high when (only one, all) of the inputs is a one for an AND gate.

4. The timing diagram in Figure 2–19 shows the inputs A and B for an AND gate. Draw the output C.

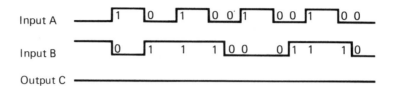

Input A

Input B

FIGURE 2–19

Output C

5. The logic symbol for the _____ is shown in Figure 2–20.

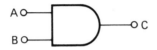

FIGURE 2–20

6. How many combinations of inputs are possible if an AND gate has four inputs?

7. How many output pins does an AND gate have?

8. Write the Boolean equation for the logic function in Figure 2–21.

FIGURE 2–21

9. What are the indeterminate or disallowed conditions or voltage levels for a TTL AND gate?

10. Look at Figure 2–22. Complete the following: If A = 1 and B = 0, then C = _____ and X = _____. If A = _____, B = _____, and C = _____, then X = 0.

FIGURE 2–22

The OR Gate

The OR gate has two or more inputs and only one output. The OR gate gives a one if any one or more of its inputs are in the one state. The OR gate concept is clearly illustrated by the circuit shown in Figure 2–23. If either switch A *or* switch B is closed (state 1), then the light L_1 will go on (state 1). The truth table in Figure 2–24 summarizes all the positions of switches A and B.

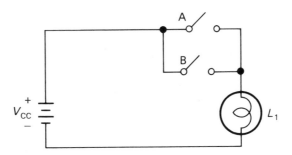

FIGURE 2-23

OR Gate Analog Circuit

A	B	L_1
0 Open	0 Open	0 Off
0 Open	1 Closed	1 On
1 Closed	0 Open	1 On
1 Closed	1 Closed	1 On

FIGURE 2-24

OR Gate Truth Table

C = A + B

FIGURE 2-25

OR Gate Logic Symbol and Boolean Equation

A + B + C + D = E

FIGURE 2-26

Four-Input OR Gate

The logic symbol and Boolean equation for an OR function are shown in Figure 2–25. The equation C = A + B is read as "C equals A OR B." The plus (+) symbol denotes the OR function and does *not* mean plus or addition. Two or more inputs are expressed in a similar fashion. For example, in Figure 2–26, the four inputs are expressed by the Boolean equation A + B + C + D = E.

The circuit in Figure 2–27 is an OR gate circuit that uses diodes. If both A and B inputs are grounded, keeping the diodes reverse biased, no current will flow. Therefore, no voltage will be dropped across resistor R_1; point C will remain low. If a high ($+V_{CC}$) is applied to A or B, current will flow through that diode and R_1 to ground. The resultant voltage drop across R_1 brings the output C high (V_{CC} minus the diode drop).

The output waveforms of an OR gate with two different pulse trains are shown in the timing diagrams in Figure 2–28. The first output waveform is done for you. Try to complete the second one yourself. Follow the input waveforms for each change as you did for the AND gate and record the outputs. The correct timing diagram is given at the bottom of Figure 2–28.

Figure 2–29 shows another OR gate circuit. Here, the circuit uses transistors. Try to follow the action. Remember, it only takes one path for current to flow to create a one state at output D. This state is caused by the current flow through R_1.

STOP Do Experiment 2-2

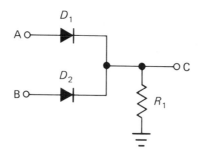

FIGURE 2-27

OR Gate Circuit Using Diodes

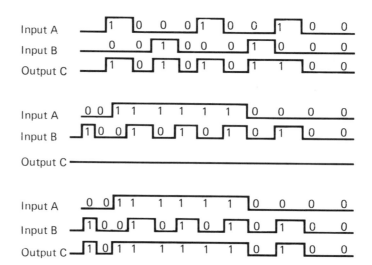

FIGURE 2-28

OR Gate Timing Diagrams

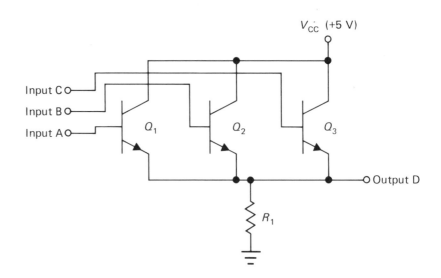

FIGURE 2-29

OR Gate Analog Circuit Using Transistors

SELF-TEST EXERCISE 2-3

1. The OR gate is a (decision-making, memory) element.
2. The OR gate will give a one output when (one or more, none) of the inputs is high.

3. The _____ gate logic symbol is shown in Figure 2–30.

FIGURE 2–30

4. Given the OR gate input signals in Figure 2–31, draw the output timing diagram.

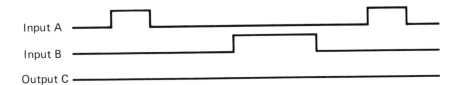

FIGURE 2–31

5. An OR gate can have two or more input pins and _____ output pin(s).

6. Write the Boolean equation for the logic function in Figure 2–32.

FIGURE 2–32

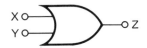

7. The truth table in Figure 2–33 is for the _____ gate.

A	B	C
0	0	0
0	1	1
1	0	1
1	1	1

FIGURE 2–33

8. Write the Boolean equation for the logic function in Figure 2–34.

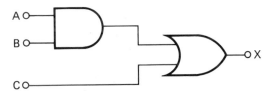

FIGURE 2–34

9. Identify the following Boolean equations by logic function—for example, inverter, OR, or AND gate:
   a. $A \cdot B \cdot C = D$
   b. $A = \overline{B}$
   c. $C + D = K$
   d. $G + E + F = Z$

The NAND Gate

The logic symbol and truth table for the NAND gate are shown in Figure 2–35. The Boolean equation for the $\overline{\text{NAND}}$ gate is a combination of the AND and inverter equations: $C = \overline{A \cdot B}$. This equation is read as "C equals A AND B NOT." Note in the truth table for the NAND gate the difference between AND and NAND: the NAND gate operates like an AND gate with the output inverted.

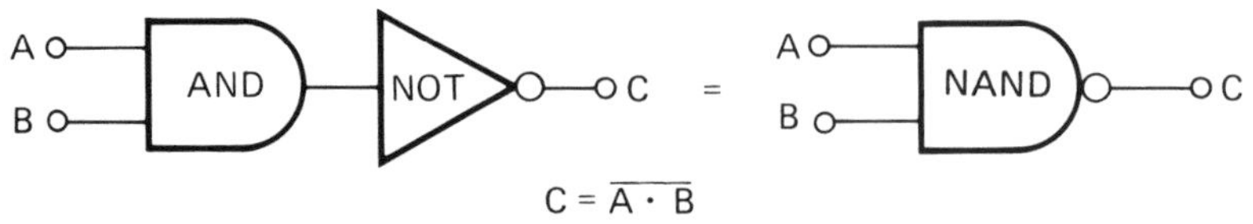

$$C = \overline{A \cdot B}$$

A	B	AND	NAND
0	0	0	1
0	1	0	1
1	0	0	1
1	1	1	0

FIGURE 2–35

NAND Gate Logic Symbol and Truth Table

The NOR Gate

Just as the NAND gate means NOT–AND, the NOR gate means NOT–OR. A NOR gate is an OR gate followed by an inverter. Figure 2–36 shows the symbolic representation and truth table of a NOR gate. The Boolean equation for the NOR gate is a combination of the OR and inverter equations: $C = \overline{A + B}$. This equation is read as "C equals A OR B NOT."

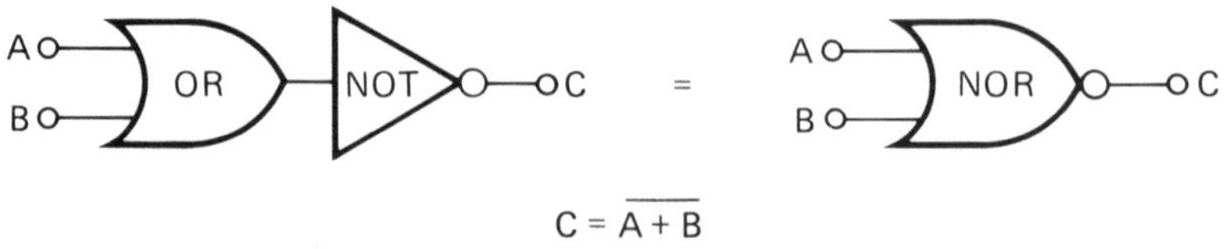

$$C = \overline{A + B}$$

A	B	OR	NOR
0	0	0	1
0	1	1	0
1	0	1	0
1	1	1	0

FIGURE 2–36

NOR Gate Logic Symbol and Truth Table

NAND AND NOR GATES AS INVERTERS

The NAND and NOR gates can be used to function as any one of the basic elements. A NAND or NOR gate with all the inputs tied together is equivalent to an inverter. See Figure 2–37. During the design of a logic system, an engineer may have gates on a chip that are not being used. To be cost efficient and to avoid confusion by using the minimum number of gates, a logic gate on an existing chip can be used in the manner just described. Note that it is a good practice to number the logic symbols on a schematic by chip—for example, "U_1 1/6", as we did in Experiment 2–1. This practice makes it easier to recognize gates on a chip that are not being used.

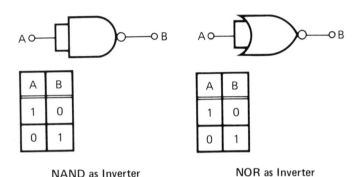

A	B
1	0
0	1

NAND as Inverter

A	B
1	0
0	1

NOR as Inverter

FIGURE 2–37

NAND and NOR Gates as Inverters

EXCLUSIVE OR GATES

While the OR gate is enabled (goes high) when either one of or all of the inputs are at logic one, the output of the Exclusive OR gate goes high only when one of the inputs is high (not when both are high). It is low when both inputs are low. The Exclusive OR logic symbol and truth table are shown in Figure 2–38. The Boolean equation for the Exclusive OR function is $C = A \oplus B$. The plus sign with a circle around it indicates the Exclusive OR function. The Exclusive OR gate will be discussed more thoroughly in a later chapter.

$C = A \oplus B$

A	B	C
0	0	0
0	1	1
1	0	1
1	1	0

FIGURE 2–38

Exclusive OR Gate Logic Symbol and Truth Table

NAND AND NOR USED FOR AND AND OR

We know from an earlier discussion that NAND logic functions can be achieved by adding an inverter to the AND gate. Similarly, a NOR function can be achieved from an OR gate. We also know that a NAND gate can be used as an AND gate if we invert the output. The <u>Boolean</u> equation for a NAND inverted to an AND is written as C = $\overline{\overline{A \cdot B}}$. As in any equation, the two negatives cancel out, which leaves C = A · B, the AND function. The same is <u>true if</u> we invert a NOR for an OR. The Boolean equation would be C = $\overline{\overline{A + B}}$, which simplifies to C = A + B. See Figure 2–39.

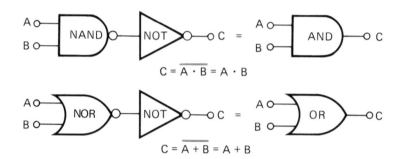

$$C = \overline{\overline{A \cdot B}} = A \cdot B$$

$$C = \overline{\overline{A + B}} = A + B$$

FIGURE 2–39

NAND and NOR Gates Used
for AND and OR Gates

OR AND AND FROM NAND AND NOR

The NAND gate truth table is shown in Figure 2–40. The inputs to the NAND gate are X and Y. Inverters are added to both inputs. The two inputs to these inverters are A and B. The incoming signals A and B are inverted and sent to the NAND gate inputs X and Y. Assume a circuit formed with the inverters and the NAND gate as a single function. A and B are inputs; C is the output. See Figure 2–40. We find the NAND gate is equivalent to the OR function when we compare it to the OR gate truth table.

An AND gate may be formed by adding inverters, in the same way as just described, to the input of a NOR gate. We then find the NOR gate truth table to be equivalent to the AND gate truth table, as shown in Figure 2–41.

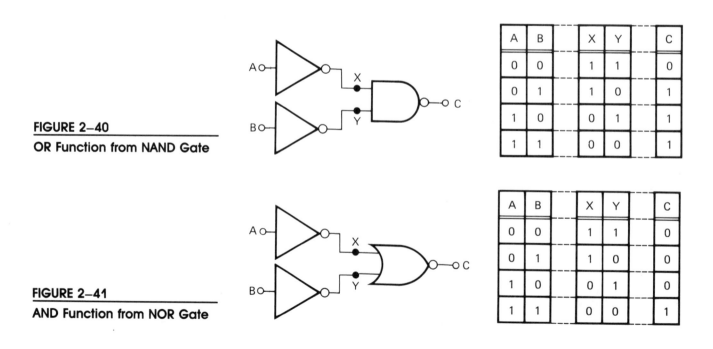

FIGURE 2–40

OR Function from NAND Gate

A	B	X	Y	C
0	0	1	1	0
0	1	1	0	1
1	0	0	1	1
1	1	0	0	1

FIGURE 2–41

AND Function from NOR Gate

A	B	X	Y	C
0	0	1	1	0
0	1	1	0	0
1	0	0	1	0
1	1	0	0	1

SELF-TEST
EXERCISE 2-4

1. A _____ gate is an AND gate followed by an inverter.
2. A _____ gate is an OR gate followed by an inverter.
3. The output of a NAND gate is the complement of the output of the _____ gate.
4. NOT-AND means _____.
5. Tying together the inputs of a NAND or a NOR gate creates the _____ gate.
6. The Boolean equation for the _____ gate is $C = A \oplus B$.
7. The Exclusive OR symbol is shown in Figure 2-42. True or False? If false, draw the correct symbol.
8. When the output of a NAND gate is inverted, the _____ gate is produced.
9. The _____ gate can be created from a NAND gate with inverters on the input.
10. Never use a NAND gate with the inputs tied together. True or False?
11. Identify the following Boolean equations by logic gate function:
    a. $C = \overline{A \cdot B}$
    b. $C = A + B$
    c. $X = \overline{Y}$
    d. $Y = \overline{X \cdot Z}$
    e. $C = \overline{A + B + C}$
    f. $D = A \oplus C$

FIGURE 2-42

PRACTICAL LOGIC CIRCUITS

There are many different ways of electrically or mechanically obtaining the particular logic functions required. Some of the popular ways are discussed here.

Relays and Switches

The basic AND, OR, and inverter (NOT) functions can be obtained easily by using *relays*. A typical AND gate using relays is shown in Figure 2-43. Contacts A and B are normally open (zero state). When power is applied to relays A and B, the switches close (one state) and the lamp lights (one state). Note that the lamp lights only when both relays are powered (one state), which is an AND gate function. OR and NOT gate configurations are shown in Figures 2-44 and 2-45. Analyze these relay circuits and verify them.

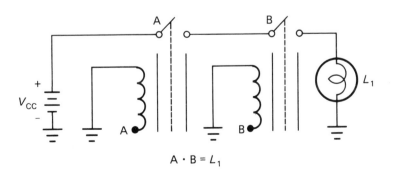

FIGURE 2-43
Relay Circuit AND Gate

$A \cdot B = L_1$

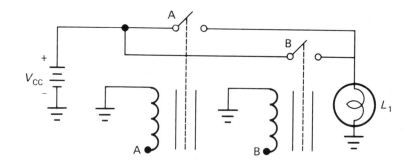

FIGURE 2-44

Relay Circuit OR Gate

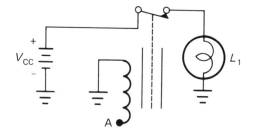

FIGURE 2-45

Relay Circuit NOT Gate

Relay switching circuits can be combined in many ways to form any logic function. Some heavy industrial controls use relay logic circuits because of their high-power capability and their reliability. In other applications, however, manually operated switches are used. Mechanical or manual switching logic is necessary, for example, in electromechanical pinballs and games. However, because this type of circuit is large, operates slowly, and consumes much power, it is not the most advantageous choice for many modern electronic applications.

 Do Experiment 2–3

Discrete Logic Circuits

Discrete logic circuits are made up of individual electronic components such as transistors, resistors, and other devices. These circuits are used primarily in high-power applications and, even then, are being replaced with logic functions that are implemented with integrated circuits.

Integrated Circuits

Integrated circuits (ICs) are semiconductor devices that have circuits in miniature form on a single chip. They are small in size (one tenth of an inch square), very low in cost (because they can be easily mass-produced), and very low in power consumption. Furthermore, no connecting components are involved; a complete circuit exists on a single chip.

Integrated circuits have been in existence for over twenty years. Their complexity has increased enormously in this time, but their cost has decreased. Not only can the basic logic gates be integrated, but the large sequential or combinational circuits also can be.

Integrated circuits that implement simple logic functions like NAND, NOR, AND, and OR and that involve less than twelve gates are known as *small-scale integration* (SSI) chips. Complete functional circuits are called *medium-scale integration* (MSI) chips and have twelve to 100 gates. Complete circuits integrated on a single chip, as in a computer, are referred to as *large-scale integration* (LSI) chips. They involve ICs with more than 100 but less than 1000 gates. *Very large-scale integration* (VLSI), *ultra large-scale integration* (ULSI), and *super large-scale integration* (SLSI) are terms often used, but not yet recognized, as standard. They describe ICs with 1000 or more gates.

Although there are thousands of ICs on the market, they can be grouped into a relatively small number of families. Methods of fabrication, supply voltage requirements, and factors such as susceptibility to noise are fairly uniform within a family, but the differences between families are considerable. Before we discuss the various families, however, we must become familiar with the terms associated with integrated circuits.

Fan-out.   *Fan-out* refers to the number of standard gate inputs that can be driven by a single IC output. A gate with a fan-out of 5 can drive five standard inputs.

Interface Circuitry.   Whenever circuits of one IC family are connected to circuits based on a different family, problems occur that make additional circuitry necessary. This circuitry is called *interface circuitry*.

Power Dissipation.   *Power dissipation* is the amount of power dissipated in heat plus the power required by the gate or system of gates to operate.

Noise.   *Noise* is any unwanted electrical disturbance or signal that affects the operation of an IC. This noise can cause a gate to "see" a signal or pulse that is not a desired part of the circuit.

Noise Immunity.   *Noise immunity* is the amount of noise that can be applied to an input (high or low) without causing a gate to produce an incorrect output.

Propagation Delay.   *Propagation delay* (t_p) refers to the time required for a change at the input of the logic circuit to result in a stable change at the output with no error. In some cases, the propagation delay can be different for a high to a low change ($t_{P_{HL}}$) and the corresponding low to high ($t_{P_{LH}}$).

Look at the inverter gate waveforms in Figure 2–46. These are the actual signals you would see on an oscilloscope. Note the slight delay before the output responds to the change in input. This delay is the propagation delay (t_p). It is usually measured from the 50% level of the input leading edge to the 50% level of the output leading edge.

Rise Time and Fall Time.   By adjusting the time band on an oscilloscope, you would see that the square wave in Figure 2–46 is not truly square. The *rise time* (t_r) is the time it takes the pulse voltage to rise from

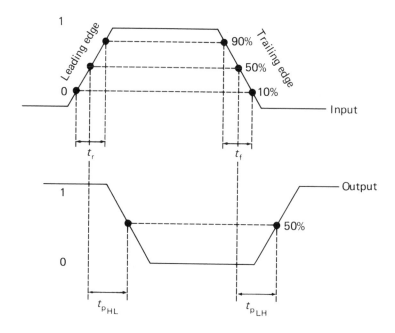

FIGURE 2–46

Square Wave

10% to 90% of its maximum value. The *fall time* (t_f) is the time it takes for the pulse voltage to fall from 90% to 10% of its maximum value.

FAMILIES OF INTEGRATED CIRCUITS

TTL Family

There are two basic series of *transistor–transistor logic* (TTL): the 5400 series, which is principally for military use because of its better temperature characteristic, and the 7400 series, which is for commercial use. While any 7400 series IC can be replaced with an equivalent 5400J series IC, a 5400 series should *not* be replaced with a 7400 series since temperature extremes may cause failure. Also, since the pin-out may be different from that of a 5400J series, it is wise to check a data book for equivalent specifications.

TTL technology is most widely used for small-scale and medium-scale integrated circuits. Sub-families of TTL ICs are divided primarily on the basis of speed and power specifications. These sub-families (standard TTL, high-speed TTL, low-power TTL, Schottky-clamped TTL, and low-power Schottky TTL) will be described after we define some new terms:

1. *Power supply voltage* (V_{CC}): Standard specified supply voltage of +5.0 V ± 5% (± 10% for 5400 series).
2. *High-level input voltage* (V_{IH}): Voltage required for logic one at an input; guaranteed minimum of +2.0 V.
3. *Low-level input voltage* (V_{IL}): Voltage required for a logic zero at an input; guaranteed maximum of +0.8 V.
4. *High-level output voltage* (V_{OH}): Voltage level output from an output in the logic one state; guaranteed minimum of +2.4 V.
5. *Low-level output voltage* (V_{OL}): Voltage level output from an output in the logic zero state; guaranteed maximum of +0.4 V.
6. *High-level input current* (I_{IH}): Current flowing into an input when a logic one voltage is applied to that input.

7. *Low-level input current* (I_{IL}): Current flowing from an input when a logic zero voltage is applied to that input.
8. *High-level output current* (I_{OH}): Current flowing from the output while the output voltage is at logic one.
9. *Low-level output current* (I_{OL}): Current flowing into an output while the output voltage is at logic zero.
10. *Logic high state for TTL:*
   —V_{IH} must be $+2.0$ V or greater.
   —I_{IH} will not exceed 40 μA.
   —V_{OH} will be $+2.4$ V or greater.
   —I_{OH} will source at least 400 μA.
11. *Logic low state for TTL:*
   —V_{IL} must not exceed $+0.8$ V.
   —I_{IL} will source up to 1.6 mA.
   —V_{OL} will not exceed $+0.4$ V.
   —I_{OL} will sink at least 16 mA.

Standard TTL.   Properties common to basic or standard TTL include a supply voltage (V_{CC}) of $+5$ V, a typical gate delay of 10 ns, a 12 mW gate dissipation, a 35 MHz maximum operating frequency, a noise margin of 0.8 V, a fan-out per input of 1, and a fan-out per output of 10.

High-Speed (High-Power) TTL.   With a 6 ns delay time, an operating speed of 50 MHz, power consumption of 22 mW, and very good noise immunity, these devices are, however, being replaced with faster Schottky TTL gates. The letter H designates a device as high-speed; for example, a 74H04 is a high-speed 7404. Note that a standard TTL gate can only fan-out to seven high-speed TTL gates.

Low-Power TTL.   Low-power TTL devices have a 1 mW power dissipation, are slowed to a 33 ns propagation delay time and only 3 MHz, but exhibit very good noise immunity. One low-power TTL gate will fan-out to ten low-power TTL gates or one standard TTL input. The letter L designates a device as low-power; for example, a 74L04 is a low-power 7404. Most of these devices are being replaced with CMOS devices in newer designs. (Making a direct replacement in existing equipment with CMOS is not advisable since CMOS is a different family and usually requires interface circuitry.)

Schottky-Clamped TTL.   Schottky-clamped TTL devices are very fast, with only a 3 ns propagation delay time. They exhibit good noise immunity, a 125 MHz clock frequency, and a 19 mW power dissipation. They are identified by the letter S, as in a 74S04.

Low-Power Schottky TTL.   Low-power Schottky TTL devices are characterized by a 2 mW power dissipation, a 9.5 ns propagation delay time at 45 MHz maximum frequency, and good noise immunity. They fan-out to ten of the same family of low-power Schottky. They are identified by the letters LS, as in a 74LS04.

REPLACEMENT GUIDE

Whenever possible, replace a chip with one from the same sub-family. Adhere to the following guidelines:

—Replace H (high-speed) designation with H or S designation. Do not use LS.

—Replace L (low-power) designation with L or LS designation. Replacing with other types may cause failure of circuits inputting to this one.

—Replace S (Schottky) designation with S designation only.

—Replace LS (low-power Schottky) designation with LS designation only.

If no letter designation is on a chip to be replaced, use a standard TTL chip. A high-speed (H) or Schottky (S) TTL chip can be used if power dissipation is not a problem. *Note:* In some cases, chips from other subfamilies can be used for replacement. However, the operation of the circuit must be studied carefully before this kind of replacement is done so that the possibility of damaging other circuit components is eliminated.

ECL Family

Emitter-coupled logic (ECL), also referred to as *current-mode logic*, provides gates that operate at higher speed than TTL circuits. The high speed and high cost of ECL circuits limit most of their use to large and expensive computers. There are three types of ECLs, all of which use a supply voltage of -5.2 V $\pm$ 10% and have a fair noise immunity. The three types and their corresponding characteristics are as follows:

1. ECL with 4 ns delays: Fan-out to 25 inputs, power dissipation of 22 mW, 4 ns propagation delay time, 70 MHz operating frequency.
2. ECL with 2 ns delays: Fan-out to 25 inputs or 50 Ω, power dissipation of 25 mW plus load, 2 ns propagation delay time, 125 MHz operating frequency.
3. ECL with 1 ns delays: Fan-out to 10 low-impedance inputs or 50 Ω, power dissipation of 60 mW plus load, 1 ns propagation delay time, 400 MHz operating frequency.

CMOS Family

Complementary metal-oxide semiconductor (CMOS) devices are gaining in popularity because of their low power consumption. A 0.01 mW static state and approximately 1 mW at 1 MHz are typical power requirements. They only use appreciable power during the actual switching state.

Operating frequencies of 5 MHz make CMOS devices slower than TTL devices. The typical propagation delay time is 25 ns to 70 ns, depending on the type of CMOS used.

Supply voltages range from $+3.0$ V to $+18.0$ V. CMOS gates have very good noise immunity and generate much less switching noise than gates in other families. The fan-out to other CMOS gates is 50 or more. CMOS technology is ideal for battery-operated equipment.

RTL Family

For *resistor–transistor logic* (RTL), the fan-out is about 5 and the propagation delay time is about 30 ns. RTL gates are very sensitive to noise.

While their low cost and simple operation were advantages at one time, they are obsolete today, having been replaced with TTL gates.

DTL Family

Diode–transistor logic (DTL) gates use a supply voltage of -5 V $\pm$ 10%. Propagation delay time is about 30 ns. These gates are three times less sensitive than RTL gates. The fan-out is about 8. Increasingly, DTL gates are being replaced by TTL gates.

 Do Experiment 2–4

SELF-TEST EXERCISE 2–5

1. Relays and switches can be used to obtain different logic functions. True or False?
2. Relay controls are used in _____ applications.
3. _____ circuits are made up of individual components.
4. _____ are miniature circuits on a single chip.
5. LSI means _____ _____ _____.
6. Propagation delay is the time needed for a technician to replace an IC. True or False?
7. _____ is the number of inputs that can be safely connected to an output.
8. (ECL, TTL, CMOS) can be used where very high speed is required.
9. _____ and _____ are obsolete IC families today.
10. Noise is a desired operation of an integrated circuit. True or False?
11. The ECL family of logic circuits is (cheaper, more expensive) than the TTL family of circuits.
12. The time for a signal to go from 10% to 90% of its maximum value is _____.
13. TTL is always powered at _____ V, V_{CC}.
14. A _____ series TTL is the military version.
15. A Schottky TTL gate can replace the high-speed TTL gate in any circuit. True or False?
16. CMOS is (slower, faster) than TTL.
17. CMOS can be powered at $+5$ V or $+12$ V, V_{CC}. True or False?
18. _____ technology is used for most battery-operated equipment.
19. If one standard TTL can fan-out to ten standard TTLs, and if one LS TTL can fan-out to only one standard TTL, then one standard TTL can fan-out to how many LS TTL chips?

SUMMARY

The two basic types of logic circuits used in digital equipment are the AND gate and the OR gate. An AND gate gives a high output when all inputs are high. An OR gate gives a high output when any one or more inputs are high. An Exclusive OR gate is high only when one of the inputs is high, not when both or neither are high.

An inverter gate is used to "invert" the input. The output is always the complement of the input. A NAND gate function is the same as an

AND gate function except that the output is inverted. A NOR gate function is the same as an OR gate function except that the output is inverted.

The purpose of refreshing the data stored in a memory cell is to restore or recharge it. When a signal is repeatedly sent to some devices (even our eyes), it appears constant.

A high for TTL positive logic is $+2.4$ V to $+5$ V. Ground to $+0.8$ V is a low. Any voltage between $+0.8$ V and $+2.4$ V is considered an undetermined state.

Fan-out is the number of inputs that can be safely driven by an IC output. TTL ICs can fan-out to ten of the same sub-family. CMOS fan-out is 50 or more. Propagation delay is the amount of time it takes for an input signal to result in a stable change on the output.

SSI is the abbreviation for small-scale integration; MSI, for medium-scale integration; and LSI for large-scale integration. Each designation indicates the number of gates integrated in a circuit and the complexity of the IC.

CHAPTER 2
REVIEW EXERCISES

1. Draw the standard logic symbols for the AND, OR, NAND, NOR, and inverter gates.

2. Assuming positive TTL logic levels of 0 V and $+5$ V, analyze the operation of the logic circuit in Figure 2–47 and develop a truth table.

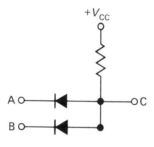
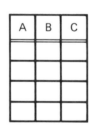

FIGURE 2–47

A	B	C

3. Assuming positive TTL logic levels of 0 V and $+5$ V, analyze the operation of the logic circuit in Figure 2–48 and develop a truth table.

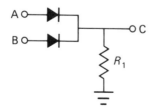
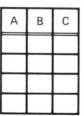

FIGURE 2–48

A	B	C

4. Write the logic function that describes each of the following logic equations:

   a. $G = \overline{N + M}$
   b. $H = J \cdot K \cdot L$
   c. $A = \overline{B}$
   d. $Q = \overline{R \cdot S}$
   e. $CLR = PB + RST$
   f. $DR = AD \oplus CB$

5. Study the logic diagram in Figure 2–49. If A = 1 and B = 1, then:

   a. What is the output at X?
   b. Draw a truth table for the circuit.
   c. What is the logic function represented?

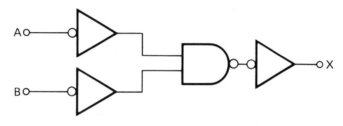

FIGURE 2–49

6. The ignition system of a car is set up so that the engine cannot be started unless the driver's front door is closed, the front seat belts are latched, and the ignition key is on. What logic function is implied? Draw the logic symbol for this function.

7. Identify each truth table in Figure 2–50 with the logic function it represents. In each case, A and B are the inputs and C is the output.

A	B	C
0	0	0
0	1	1
1	0	1
1	1	1

a._____

A	B	C
0	0	1
0	1	0
1	0	0
1	1	0

b._____

A	B	C
0	0	0
0	1	0
1	0	0
1	1	1

c._____

A	B	C
0	0	1
0	1	1
1	0	1
1	1	0

d._____

A	C
0	1
1	0

e._____

FIGURE 2–50

8. What logic function is being performed by the circuit in Figure 2–51?

  a. AND
  b. OR
  c. NAND
  d. NOR

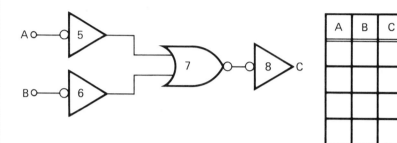

A	B	C

FIGURE 2–51

9. A home intrusion alarm system is designed to sound a bell if any one of the following conditions occurs: front or back doors open, any window opens, garage door opens. What logic function is implied? Draw the logic symbol for this function.

10. Write the logic equation for the relay circuit shown in Figure 2–52.

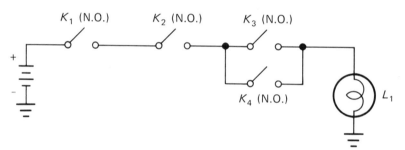

FIGURE 2–52

11. A logic circuit has five inputs. How many possible input combinations can it have?

  a. 2
  b. 4
  c. 5
  d. 16
  e. 32

12. Figure 2–53 shows logic timing diagrams for the inputs A and B. Draw the output waveforms you would expect for an AND gate and for an OR gate.

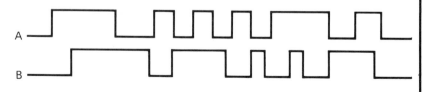

$A \cdot B =$ _____

FIGURE 2–53

$A + B =$ _____

13. The symbol in Figure 2–54 is an inverter symbol. True or False?

FIGURE 2–54

14. Which of the following equations indicate the AND function?

    a. $A \cdot B$
    b. $A \times B$
    c. $A \oplus B$
    d. $A + B$
    e. $\underline{AB}$
    f. $\overline{ABC}$
    g. $\overline{A + B}$

15. Is +1.6 V a logic low, logic high, or disallowed state for a TTL integrated circuit?

16. Can NAND gates be safely operated with their inputs tied together?

17. Draw the truth table for a two-input Exclusive OR gate.

18. When would relays and switches be used instead of integrated circuits to perform the same logic functions?

19. What does LSI stand for?

20. What does the term *fan-out* for an IC mean?

21. Which has a larger fan-out capability, CMOS or TTL?

22. What is propagation delay?

23. A 74LS00 is a safe replacement for a 7400. True or False? Explain your answer.

24. ECL is used for its high speed. True or False?

25. Why is CMOS used for battery-operated digital circuits?

26. Why do some logic probes affect circuit operation?

27. What are the two methods used to test logic systems?

28. What does *refresh* mean? Why is it used?

| EXPERIMENT 2–1 | # Inverters in Discrete and Integrated Circuits |

PURPOSE

This experiment is designed to show the inverting operation in both discrete and integrated circuit form. You will complete a truth table and draw a timing diagram for this basic operation.

EQUIPMENT

—1 dual-trace oscilloscope (Equip$_1$)
—1 logic probe (Equip$_2$)
—1 digital experimenter (Equip$_3$)
—1 SN7404 hex inverter IC (U_1)
—1 1000 Ω, 1/2 W resistor (R_1)
—1 4700 Ω, 1/2 W resistor (R_2)
—1 2N2222A NPN transistor (Q_1)

PROCEDURE: PART I

Step 1 Wire the circuit as shown in Figure 2E1–1.

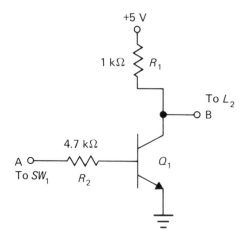

FIGURE 2E1–1

Step 2 Connect input point A to a TTL data switch (SW_1) that is in the +5 V position.

Step 3 Connect the logic probe between +5 V and ground according to the manufacturer's instructions.

Step 4 Touch the probe to point A and then to point B.

ACTIVITY

Record the results in row 1 of the truth table in Figure 2E1–2.

	Input A	Output B
Row 1		
Row 2		

+5 V = 1
GND = 0

FIGURE 2E1–2

Step 5 Move point A, data switch SW_1, to the ground position.

Step 6 Measure point A and point B with the probe.

ACTIVITY

Record the results in row 2 of the truth table in Figure 2E1–2.

Step 7 Connect a logic indicator (L_1) to point A as shown in Figure 2E1–3.

NOTE: If the digital experimenter you are using has logic indicator LEDs, they are already protected by additional circuitry for this direct connection and will glow only when this point is high. If logic indicators are not available on the experimenter you are using, use a 330 Ω, 1/2 W resistor in series with an LED to ground.

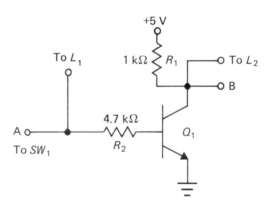

FIGURE 2E1–3

Step 8 Connect logic indicator (L_2) to point B as shown in Figure 2E1–3.

Step 9 Alternate the data switch SW_1 to the +5 V and the ground position to see the inverting action of this simple transistor inverter.

Step 10 Remove the data switch SW_1 connection at point A and then connect point A to a 1 Hz clock.

Step 11 Observe the logic indicators. You can see the output change as the input changes.

Step 12 Connect channel A of a dual-trace oscilloscope to point A of the circuit.

Step 13 Connect channel B to point B.

Step 14 Change the 1 Hz clock to 1000 Hz to get a steady signal on the scope.

ACTIVITY

In Figure 2E1–4, draw the two signals you see on the screen. This drawing is a timing diagram of the inverting circuit.

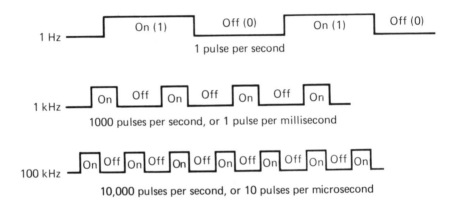

FIGURE 2E1–4

ACTIVITY

Explain why both logic indicators are now lit.

Step 15 Increase the clock rate to 100 kHz.

NOTE: Both logic indicators are still lit, but they may appear brighter. The reason that L_1 and L_2 do not appear to be flashing is that the flashing occurs at a rate that our eyes are not able to follow. The higher the flash or pulse rate, the less we can "see" the on–off state. Also, the higher the pulse rate (Hz), the shorter the on–off time; therefore, the light may appear brighter. See Figure 2E1–5.

FIGURE 2E1–5

Digital equipment, including some calculators, uses this fact to save power. Although the display on a calculator appears to be constant, it is actually flashing or pulsing at a rate you cannot see, which means it is on only half the time. A considerable savings of battery power is provided by the *refresh circuit* that supplies this on–off display. Refresh circuits are widely used, particularly in microcomputer memory circuits.

> **Check:** What is the logic function performed by the previous circuit?
>
> inverter

> **Check:** When you applied the clock, you saw that the input and the output are always _____.
>
> inverted

PROCEDURE: PART II

You will use a TTL IC in the following steps to see how an inverter operates. Fourteen and sixteen pin versions are available. A notch or a dot is used to designate pin 1.

NOTE: Usually, in a chip diagram like the one in Figure 2E1–6, only the logic symbols, pin numbers, and IC number will be shown. Check ground and V_{CC} supply connections; they vary from one type of IC to another.

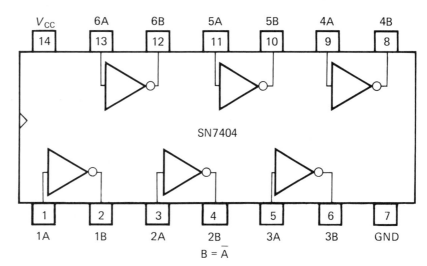

FIGURE 2E1–6

Step 1 Install the SN7404 IC on the breadboard socket available on the digital experimenter you are using.

Step 2 Apply power to the IC by connecting pin 14 to +5 V and pin 7 to ground.

NOTE: The connections for V_{CC} pin 14 and ground pin 7 are shown on the schematic in Figure 2E1–7. These connections are sometimes shown on schematics for the first logic gate on a chip but are seldom shown for every gate on the chip.

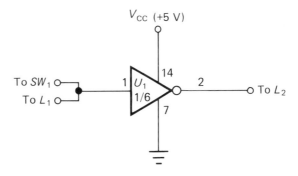

FIGURE 2E1–7

Step 3 Connect one of the inverters as shown in Figure 2E1–7.

NOTE: In Figure 2E1–7, U_1 identifies the SN7404 hex inverter listed under "EQUIPMENT" at the beginning of this experiment. The 1/6 indicates that this gate is the first of six logic gates on the chip. Succeeding gates used on a schematic are marked U_1 2/6, U_1 3/6, and so on. The input comes from data switch SW_1; the output is connected to logic indicator L_2. Logic indicator L_1 shows the state of the input.

Step 4 Supply a ground (0 V, L_1 off) to the input point A through data switch SW_1 and then check the output point B.

Step 5 Supply +5 V (L_1 on) to the input point A through data switch SW_1 and then check the output.

Step 6 Remove data switch SW_1 at pin 1 and leave it open.

Step 7 Check the output. L_1 will be on, and L_2 will be off.

> **Check:** An open input has the same effect as a binary _____.
>
> one

NOTE: It is not good design practice to leave inputs open, since they can pick up noise or stray signals and cause circuit problems. They should be connected to high (V_{CC}) or low (ground). Do not let the fact that logic indicator L_2 may not light if no connection (L_1) is made in the experiment confuse you. You can measure the output point B with your meter to check the voltage. It just cannot supply the current drive for L_2. Here we have another example of what poor design (leaving an input open) can do.

Step 8 Connect four inverters in series as illustrated in Figure 2E1–8.

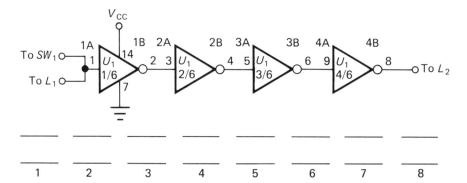

1	2	3	4	5	6	7	8

Step 9 Connect data switch SW_1 to point A and L_1 and then connect the output to L_2.

Step 10 Set data switch SW_1 to zero (L_1 off) and then to one.

Record the outputs. Record each input and output for the four gates in the blanks in Figure 2E1–8. Use the logic probe to measure them.

NOTE: An IC can have one or more inverters on a chip that do not function correctly. Do not discard the chip; just use the gates that work.

Step 11 Connect three gates in series as shown in Figure 2E1–9 and repeat the above procedure.

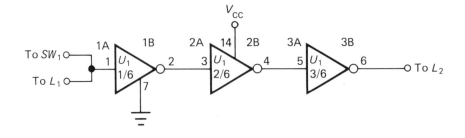

Check: Connecting an even number of gates (will, will not) invert the input.

will not

Check: Connecting an odd number of gates gives (no inversion, inverted) output with respect to the input.

inverted

EXPERIMENT 2–2 | AND and OR Gate Discrete Circuits

PURPOSE

This experiment is designed to familiarize you with the analog AND and OR gate operation. You will draw timing diagrams and complete a truth table for each gate.

EQUIPMENT

—1 logic probe (Equip$_1$)
—1 dual-trace oscilloscope (Equip$_2$)
—1 digital experimenter (Equip$_3$)
—2 1N4148 fast-switching diodes (D_1, D_2)
—1 1000 Ω, 1/2 W resistor (R_1)

PROCEDURE

Step 1 Wire the circuit as shown in Figure 2E2–1. Inputs A and B come from data switches SW_1 and SW_2.

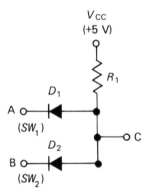

FIGURE 2E2–1

Step 2 Apply inputs to the logic gate using SW_1 and SW_2 as shown in the truth table in Figure 2E2–2 and monitor the inputs and output with the logic probe.

A	B	C
0	0	
0	1	
1	0	
1	1	

FIGURE 2E2–2

ACTIVITY

Record the results in Figure 2E2–2.

Check: What type of logic gate does this truth table represent?

AND gate

56

Step 3 Disconnect SW_2, connect this point to a 1 Hz clock (CLK) and to logic indicator L_1.

Step 4 Connect point C to logic indicator L_2 as shown in Figure 2E2–3.

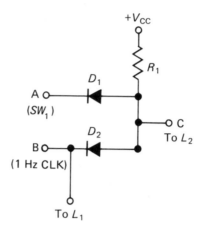

FIGURE 2E2–3

Step 5 Set SW_1 to ground position and observe L_2.

Check: When SW_1 = low, what is the output at point C? Explain your answer.

low

NOTE: The timing diagram for the preceding operation is shown in Figure 2E2–4.

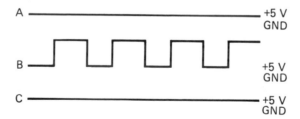

FIGURE 2E2–4

Step 6 Set SW_1 to the +5 V position and observe logic indicator L_2.

Check: When SW_1 = high, what is the output at point C? Explain your answer.

1 Hz clock

ACTIVITY

Draw the timing diagram for the preceding operation in Figure 2E2–5.

A ——————————— +5 V
                GND

B ——————————— +5 V
                GND

C ——————————— +5 V
                GND

FIGURE 2E2–5

Step 7 Disconnect SW_1 and connect this point to $\overline{CLK}$.

NOTE: $\overline{CLK}$ is 180° out of phase with CLK. This condition can be shown if you connect channel A of the scope to CLK and channel B of the scope to $\overline{CLK}$. The frequency of the clock may have to be increased to get a good signal on the scope.

ACTIVITY

Observe logic indicator L_2 in Figure 2E2–6. Explain what is taking place by drawing the timing diagram of both inputs and output in Figure 2E2–7.

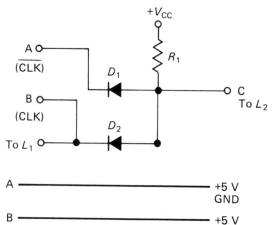

FIGURE 2E2–6

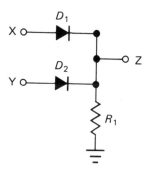

FIGURE 2E2–7

Step 8 Connect the circuit as shown in Figure 2E2–8. SW_1 and SW_2 will supply inputs X and Y. Output Z is seen with a logic probe.

FIGURE 2E2–8

Step 9 Apply voltages to X and Y as shown in the truth table in Figure 2E2–9.

X	Y	C
0	0	
0	1	
1	0	
1	1	

FIGURE 2E2–9

ACTIVITY

Check inputs and output with a logic probe and record the output in Figure 2E2–9.

> **Check:** What type of logic gate does this truth table represent?
>
> OR gate

Step 10 Connect the circuit as shown in Figure 2E2–10.

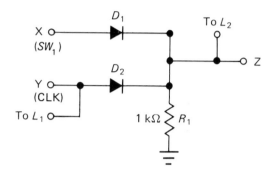

FIGURE 2E2–10

ACTIVITY

Explain this circuit operation and draw the timing diagrams for SW_1 low in Figure 2E2–11 and for SW_2 high in Figure 2E2–12.

```
X ———————————————————— +5 V
                        GND
Y ———————————————————— +5 V
                        GND
Z ———————————————————— +5 V
                        GND
```

FIGURE 2E2–11

```
X ———————————————————— +5 V
                        GND
Y ———————————————————— +5 V
                        GND
Z ———————————————————— +5 V
                        GND
```

FIGURE 2E2–12

NOTE: Since X is low, the output can go high only when the Y input goes high; thus, Z = Y in Figure 2E2–11. If SW_1 is high, the OR function means Z will always be high, as in Figure 2E2–12.

Step 11 Disconnect SW_1 and connect the X input to $\overline{CLK}$ to determine the output Z.

ACTIVITY

Draw the timing diagram in Figure 2E2–13.

```
X ———————————————————— +5 V
                        GND
Y ———————————————————— +5 V
                        GND
Z ———————————————————— +5 V
                        GND
```

FIGURE 2E2–13

NOTE: Either X or Y is always high; therefore, the output is always high.

EXPERIMENT 2–3 | AND and OR Gate Integrated Circuits

PURPOSE

This experiment is designed to show the operation of the AND and OR gate integrated circuit. You will complete a truth table and draw timing diagrams for each circuit.

EQUIPMENT

—1 logic probe with pulse indicator ($Equip_1$)
—1 dual-trace oscilloscope ($Equip_2$)
—1 digital experimenter ($Equip_3$)
—1 SN7408 IC AND gate (U_1)
—1 SN7432 IC OR gate (U_2)

PROCEDURE

Step 1 Apply power to the IC in Figure 2E3–1 by connecting pin 14 to +5 V (V_{CC}) and pin 7 to ground.

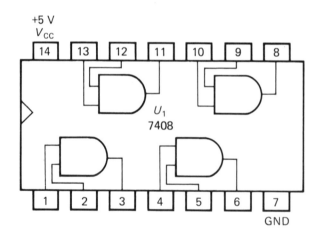

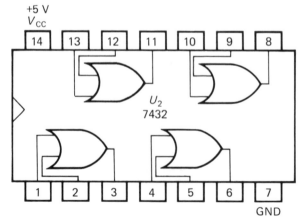

FIGURE 2E3–1

Step 2 Connect SW_1 to input pin of U_1 and connect SW_2 to input pin 2 of U_1. Connect output pin 3 to logic indicator L_1 of the experimenter.

ACTIVITY

Read the inputs and output of U_1 using the logic probe and complete the truth table in Figure 2E3–2.

A	B	C
0	0	
0	1	
1	0	
1	1	

FIGURE 2E3–2

> **Check:** What type of logic gate does this truth table represent?
>
> AND gate

NOTE: Logic indicator L_1 will give the same result as the logic probe.

Step 3 Disconnect pin 1 from SW_1 and connect pin 1 to a 1 Hz clock. Observe what L_1 is doing when SW_2 is high.

Step 4 Change the clock speed to 1000 Hz and move SW_2 to the high position.

ACTIVITY

Explain why L_1 is always on, but dimly lit.

NOTE: Use the scope to see the inputs and outputs of any circuit if it will help you to understand the relationship of one signal to another. You will also become familiar with the appearance of the actual signals that you see when you begin to troubleshoot equipment. Look at the output pin 3 using a logic probe with a pulse LED indicator. You will see that the high and low LEDs are lit. The pulse indicator should be flashing. Your logic probe may respond in a different way. Be sure to read the instructions that came with your probe.

Step 5 Change the clock speed to 100 kHz and observe L_1.

Step 6 Reduce the clock speed to 1 Hz.

Step 7 Disconnect SW_2 at pin 2 and connect pin 2 to $\overline{CLK}$. Observe the output at L_1.

ACTIVITY

Draw the timing diagram for this circuit in Figure 2E3–3.

Pin 1
CLK _____

Pin 2
$\overline{CLK}$ _____

Pin 3
Output _____

FIGURE 2E3–3

Step 8 Repeat steps 1 and 2 using U_2.

ACTIVITY

Read the inputs and output to fill in the truth table in Figure 2E3–4 for U_2 as you did for U_1.

A	B	C
0	0	
0	1	
1	0	
1	1	

FIGURE 2E3–4

Step 9 Disconnect SW_1 from pin 1 and connect pin 1 to the 1 Hz clock.

Step 10 Move SW_2 to the high position.

ACTIVITY

Explain what L_1 is doing when SW_2 is high. Explain why.

Step 11 Move SW_2 to the low position.

ACTIVITY

Explain what L_1 is doing when SW_2 is low. Explain why.

Step 12 Change the clock speed to 100 kHz and move SW_2 to the high position.

NOTE: The operation should not change; it is the same as it was when the clock was at 1000 Hz.

Step 13 Disconnect SW_2 at pin 2 and connect pin 2 to $\overline{CLK}$ to determine the output at L_1.

ACTIVITY

Draw the timing diagram for this circuit in Figure 2E3–5.

Pin 1
CLK ——————————————————

Pin 2
$\overline{CLK}$ ——————————————————

FIGURE 2E3–5

Pin 3
Output ——————————————————

EXPERIMENT 2–4

NAND and NOR Gate Integrated Circuits

PURPOSE

This experiment is designed to illustrate the operation of the NAND and NOR gate integrated circuit. You will complete a truth table and draw timing diagrams for each operation. You will also connect each gate for use as an inverter.

EQUIPMENT

—1 logic probe with pulse indicator (Equip$_1$)
—1 dual-trace oscilloscope (Equip$_2$)
—1 digital experimenter (Equip$_3$)
—1 SN7400 IC NAND gate (U_1)
—1 SN7402 IC NOR gate (U_2)

PROCEDURE

Step 1 Connect SW_1 to input pin 1 of U_1 and connect SW_2 to input pin 2 of U_1. Connect output pin 3 to logic indicator L_1 of the experimenter. See Figure 2E4–1.

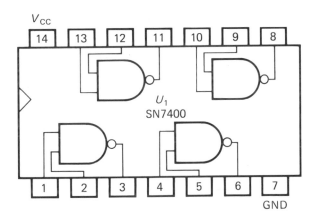

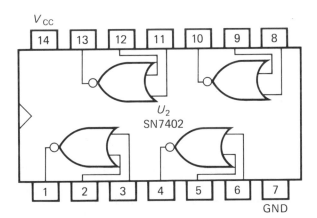

FIGURE 2E4–1

NOTE: Do not forget to apply power, +5 V (V_{CC}) pin 14 and ground pin 7. You will not always be reminded of this step in future experiments because the first step in all troubleshooting is to check the power supply.

Step 2 Apply inputs to the logic gate using SW_1 and SW_2 as shown in the truth table in Figure 2E4–2.

A	B	C
0	0	
0	1	
1	0	
1	1	

FIGURE 2E4–2

ACTIVITY

Read the inputs and output of U_1 with the logic probe and complete Figure 2E4–2.

Check: What type of logic gate does this truth table represent?

ǝʇɐƃ ᗡN∀N

NOTE: Logic indicator L_1 will give the same result as the logic probe.

Step 3 Disconnect SW_1 from pin 1 and connect pin 1 to pin 2.

Step 4 Move SW_2 high and observe the output.

Step 5 Move SW_2 low and observe the output.

Check: What logic function do the procedures in steps 3 and 4 indicate?

ɹǝʇɹǝʌuı

Step 6 Disconnect pin 1 from pin 2 and connect pin 1 to the 1 Hz clock.

Step 7 Move SW_2 high.

ACTIVITY

Explain what L_1 is doing when SW_2 is high.

Step 8 Move SW_2 low.

ACTIVITY

Explain what L_1 is doing when SW_2 is low.

NOTE: If you cannot explain what is happening, read Chapter 2 again.

Step 9 Disconnect SW_2 and connect pin 2 to $\overline{CLK}$.

ACTIVITY

Draw the timing diagram for this circuit in Figure 2E4–3.

Pin 1
CLK ———————————————————————————

Pin 2
$\overline{\text{CLK}}$ ———————————————————————————

FIGURE 2E4–3

Pin 3
Output ———————————————————————————

Step 10 Repeat the procedures in steps 1 and 2 using U_2.

NOTE: Inputs are now pin 2 and pin 3; output is pin 1. See the chip diagram (Figure 2E4–1) at the beginning of this experiment.

ACTIVITY

Complete the truth table in Figure 2E4–4 and draw the timing diagram in Figure 2E4–5 for this circuit.

A	B	C
0	0	
0	1	
1	0	
1	1	

FIGURE 2E4–4

A ———————————————————————————

B ———————————————————————————

FIGURE 2E4–5

C ———————————————————————————

3 Basic Flip-Flops

OBJECTIVES

After studying this chapter you will be able to:

1. Describe the operation of a basic sequential logic memory element.
2. Describe the operation of an RS and $\overline{RS}$ flip-flop.
3. Construct a debounce circuit.
4. Describe the operation of a D-type bistable flip-flop.
5. Construct a simple 4-bit storage register.
6. Know the synchronous and asynchronous operation of a JK flip-flop.
7. Construct a divide-by-two circuit using a TTL 7476 JK flip-flop.
8. Read and put to use flip-flop truth tables.
9. Draw timing diagrams for simple flip-flop circuits.
10. Recognize leading and trailing edges of clock pulses.

INTRODUCTION

In and of themselves, basic logic gates cannot adequately serve as temporary storage or solve interfacing problems. More complex integrated circuits have been added to the digital family to perform these functions. This chapter discusses the most versatile and widely used storage circuits, or flip-flops, available today.

BASIC SEQUENTIAL LOGIC MEMORY ELEMENT

We have learned that AND and OR gates are decision-making elements that can be combined to make complex decisions. They make decisions based on the *present* inputs to the gates; as the inputs change, so do the outputs. When a holding or storage function is required, however, a device is needed that can remember *past* events. This device is a *sequential logic memory element*.

A basic memory element constructed from an OR gate is shown in Figure 3–1. A discussion of how this simple circuit operates will help you understand how more complex circuits are designed.

Assume a logic zero at input A. This condition gives a zero at output Q, which feeds back to input B. Output Q will be zero until A goes high. Output Q will go high when A goes high, which sends a high to B. Q will remain high even if A changes to a low because B is still high. It,

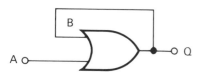

FIGURE 3–1
Basic Memory Element

66

therefore, "remembers" the high input. Output Q cannot be changed in this circuit unless Q is physically disconnected from B and both A and B are returned to logic zero.

This memory element is sufficient for any application where only one event is to be recorded in a lifetime; it is not practical in most situations. A memory element that can be *erased* or *reset* when it no longer needs to remember an event is more versatile.

A *flip-flop* (FF) is such a memory element. It can store a one or a zero as long as power remains or until it is reset. It can be recognized by its two outputs, Q and $\overline{Q}$. ($\overline{Q}$ is read as NOT Q or Q NOT.) A flip-flop is said to be in the *set* state when the output Q is high ($\overline{Q}$ will be low). A flip-flop is said to be in the *reset* state when Q is low ($\overline{Q}$ will be high).

RS FLIP-FLOPS

RS (reset–set) flip-flops can be built with NOR or NAND gates, as illustrated in Figures 3–2 and 3–3. To study the operation of an RS flip-flop, look at the NOR flip-flop in Figure 3–2. Start by setting the set input S to a high (reset R is held low). When output $\overline{Q}$ goes low, a low is applied to U_2. We have a high output with the low at R. This high feeds back to U_1 and keeps its output low even if S should now go low (remember a NOR gate needs only one high input to go low). In effect, we have crossed, or *latched*, the flip-flop—hence, the term *RS latch*. Any additional pulses to the S input of a set flip-flop will not change the output.

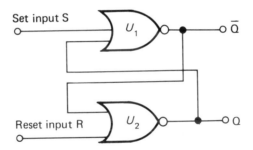

FIGURE 3–2

NOR RS Latch

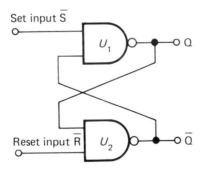

FIGURE 3–3

NAND $\overline{\text{RS}}$ Latch

When we apply a high to the reset input R (set S is held low) to reset the circuit, the output Q goes low, sending a low signal to U_1 and causing $\overline{Q}$ to go high. The circuit is reset; we have erased the memory. It will now wait for another high at S. Any additional pulses to a reset line when a flip-flop is reset will not change the output.

The outputs Q and $\overline{Q}$ will stay in the condition they were in last if both R and S are low. If both R and S go high at the same time, an unacceptable "limbo" state for the flip-flop results and should be avoided. The internal circuits that are used to build the gates are very well matched transistor circuits. They will try to switch at the exact same instant. They will race one another trying to switch the output first. Because we cannot be sure which gate will switch first, we can never be sure of the output condition. Also, in some cases, the circuit may not settle to any one output but be trapped in a *race condition*, switching back and forth. This condition should never be allowed to occur.

The logic symbol and truth table for an RS latch are shown in Figure 3–4. Learn to read flip-flop truth tables such as this one and to recognize the various symbols used in them. (Different manufacturers use different symbols; some of the most common symbols are used in this book.)

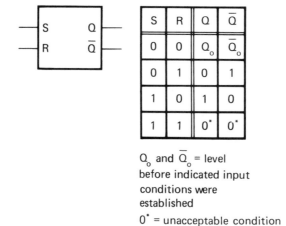

S	R	Q	$\overline{Q}$
0	0	Q_o	$\overline{Q}_o$
0	1	0	1
1	0	1	0
1	1	0^*	0^*

Q_o and $\overline{Q}_o$ = level before indicated input conditions were established

0^* = unacceptable condition

FIGURE 3–4

RS Latch Logic Symbol and Truth Table

The NAND $\overline{RS}$ latch shown in Figure 3–3 operates in a manner similar to the NOR RS latch. The difference is that the NAND gate circuit operates on logic lows instead of logic highs. That is, a low instead of a high is used to set or reset. The logic symbol and truth table for an $\overline{RS}$ latch—in this case, a 74279 TTL chip—are shown in Figure 3–5. Note that this particular IC does not have a pin-out for $\overline{Q}$ because of pin limitations.

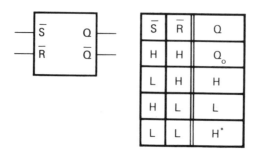

$\overline{S}$	$\overline{R}$	Q
H	H	Q_o
L	H	H
H	L	L
L	L	H^*

Q_o = level before indicated input conditions were established

H^* = unacceptable condition

FIGURE 3–5

RS Latch Logic Symbol and Truth Table

An integrated circuit, of course, does not tell you what its internal gating is. How, then, do you know whether the IC is a NAND or a NOR that uses active highs or lows? You know from the following:

1. The truth table indicates the IC's highs and lows.
2. The logic lows on a logic symbol are designated $\overline{S}$ or $\overline{R}$ (the inversion is indicated by the bar over the input).
3. The logic diagram has an open circle that indicates logic low inputs.
4. Look up the IC in a data book.

Debounce Circuits

One of the most common and useful applications of RS latch flip-flops is *switch buffering*. Push-button switches are used in digital circuits to control various functions. These switches do not make or break perfectly. The contacts bounce open and closed for a period of time. So, instead of one pulse, we get a number of pulses and the circuits are triggered many times instead of just one time. Figure 3–6 shows an example of the waveform that a switch may produce. Notice the repeated trigger pulses, or *contact bounce*.

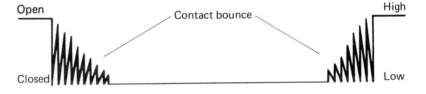

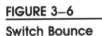

FIGURE 3–6

Switch Bounce

A typical contact bounce eliminator, or *debounce circuit*, is shown in Figure 3–7. Debounce flip-flops are used to get a single clear pulse. They operate as follows.

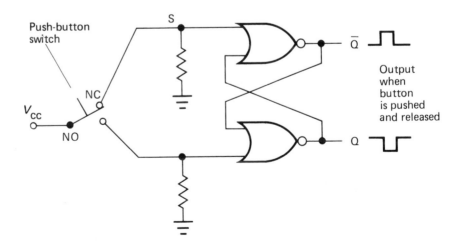

FIGURE 3–7

Debounce Circuit

The push button holds the set input high in the normally closed position. Therefore, the flip-flop is set until the switch moves from S to R, which changes the flip-flop to reset (low output at Q). When the contact returns to set, the flip-flop returns high. Since contact bounce at either input has no effect, we get a clear high-to-low output pulse.

We buffer contact bounce with a latch. If a NAND latch ($\overline{\text{RS}}$ latch) were used, the switch would be connected to ground instead of V_{cc} with the input normally in the low position. If a positive output pulse is needed with a NOR circuit (RS latch), the $\overline{Q}$ line would be used or the circuit could be held in the reset condition and a set pulse applied. Q would be a positive output. Note the output Q in the timing diagrams in Figure 3–8 for inputs R and S.

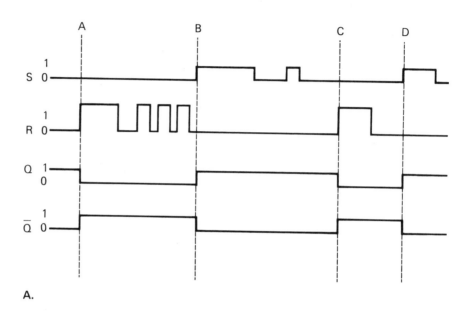

A.

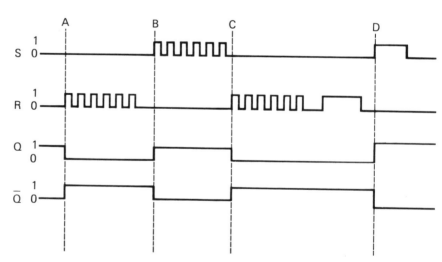

FIGURE 3–8

RS Latch Timing Diagrams

B.

We know from Figure 3–8A that the RS latch works on active highs because there is no inversion symbol to indicate active low operation. The first active high is on the reset input R at point A, which means the output Q must go low (reset). Since the previous state of Q was high, we go low at this point. Once we have reset, all additional pulses on the reset line are ignored. Therefore, we "flip-flop" to the set line to look for a high. We find one at point B. Q now goes high (set). Once the output is set, we look for another active high on the reset and ignore any additional set pulses. The next reset is at point C. Q goes low (reset). Now we look for a set pulse and find one at point D. Q goes high (set). Since there are no more reset pulses, Q stays set for the rest of the time shown on the timing diagram. $\overline{Q}$ is the complement of Q, of course.

In Figure 3–8B, the operation is exactly the same as just described. The transition points are shown for you, however.

Figure 3–9 shows $\overline{RS}$ latch timing diagrams. $\overline{RS}$ latches work on active lows, which means the output Q changes only when the set or reset goes low. Since set starts low in Figure 3–9A, the output Q must start in the set state. The next pulse we want is an active low reset, which we find at point A. Q goes low (reset). We look next for an active low on the set line. It is found at point B. Q goes high (set). The next reset is at point C. Q goes low. The next active pulse is at point D. Q sets (Q goes high). The last pulse we see is reset at point E. Q stays reset the rest of the time in this diagram. Again, $\overline{Q}$ is the complement of Q.

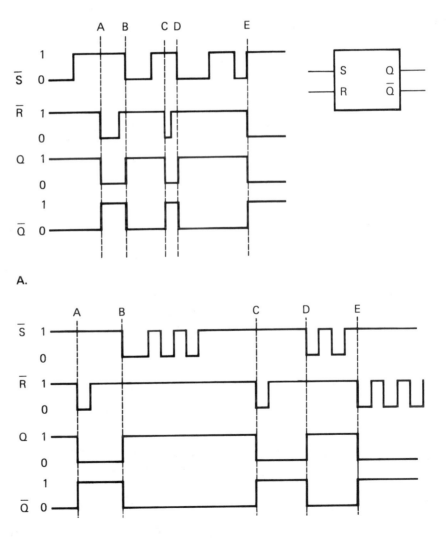

A.

B.

FIGURE 3–9

$\overline{RS}$ **Latch Timing Diagrams**

In Figure 3–9B, the operation is just the same as described for Figure 3–9A. Active lows change the output. The transition points are again marked for you.

Because RS and $\overline{\text{RS}}$ flip-flops are so easy to construct from NOR or NAND gates, they are seldom bought in IC form as a latch. Usually, they are simply wired from the IC gates where they are needed.

SELF-TEST EXERCISE 3–1

1. A flip-flop is used to temporarily store data. True or False?
2. One flip-flop can store _____ bits of data.
3. The RS flip-flop is set when Q = _____.
4. The RS flip-flop is reset when $\overline{\text{Q}}$ = _____.
5. A (low, high) applied to R will reset a NAND latch when it is set.
6. Only (lows, highs) can set or reset a NAND latch.
7. Both inputs (high, low) is an unacceptable state for a NOR latch.
8. Only (lows, highs) can set or reset a NOR latch.
9. Both inputs at (0, 1) is an unacceptable state for a NAND latch.
10. Latches are used as buffers for contact bounce. True or False?
11. The inputs should be designated by an inversion symbol when a (low, high) is used to set or reset a latch. An example is shown in Figure 3–10.

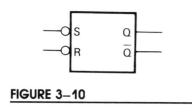

FIGURE 3–10

STOP   Do Experiment 3–1

D-TYPE FLIP-FLOPS

There are various types of D flip-flops. Like RS flip-flops, *D-type flip-flops* have two inputs and two outputs. We will examine the D-type bistable latch first.

Bistable Latches

The *bistable latch* is an element in a storage register. The logic symbol and diagram of the bistable latch are shown in Figure 3–11.

The D-type bistable latch works differently from the RS latch. D is the *data* input, where the data to be stored is applied. G is the *enable* input, which directs the flip-flop and allows the data input to be recognized or ignored. If the enable input is high, the data at D is stored in the flip-flop. Otherwise, no matter what happens at D, it is ignored. The Q output will stay in the previous stored level. As long as the enable line is high, the output will follow the data input line.

The truth table for the bistable latch and typical input and output waveforms are shown in Figure 3–12. The truth table in Figure 3–12A shows us that this latch outputs data only when the enable line is high. It is helpful, therefore, to draw lines on the timing diagram around the enable high pulse, as shown in Figure 3–12B. The data can get to Q only when these "windows" are open, and they are open only when the enable line is high. The output Q follows the data line when the enable line is high. When the enable line goes low, the output stays where it was before the enable line changed (previous state).

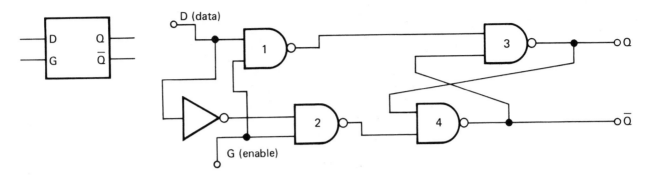

FIGURE 3–11

D-Type Bistable Latch Logic Symbol and Diagram

D	Enable	Q	$\overline{Q}$
0	1	0	1
1	1	1	0
X	0	Q_o	$\overline{Q}_o$

X = don't care
Q_o = remains in previous state
$\overline{Q}_o$ = remains in previous state

A.

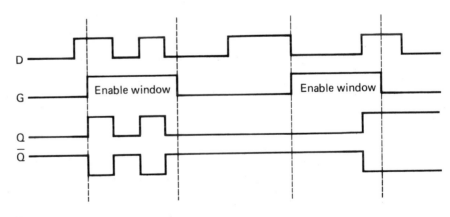

FIGURE 3–12

D-Type Bistable Latch Truth Table and Waveforms

B.

Some bistable latches work on a low or *enable not*. These latches are easily recognized because their truth table shows a bar over the enable symbol, their logic symbol carries an inversion sign, or their logic diagram has a small circle on the enable input.

Storage Registers

The most common application of bistable flip-flops is as elements in a *storage register*. In digital equipment, data is often transferred from one register to another. Figure 3–13 shows how data in the form of switches can be transferred to flip-flop registers. Notice that switch outputs are connected to the D inputs, and the enable line is connected to a common clock so that all data is transferred at the same time. Assuming that the enable line is high, the outputs will be as shown in Figure 3–13. If the enable line or clock were to go low, the outputs would be at the level of the previous switch condition. In most digital equipment, the enable line is a clock input.

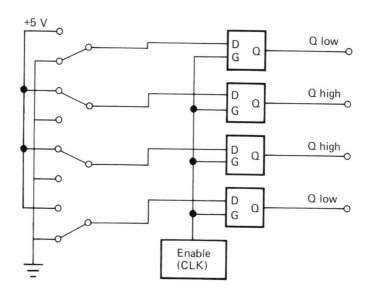

FIGURE 3–13

Storage Register

SELF-TEST EXERCISE 3–2

1. Bistable flip-flops are used as an element in (temporary storage registers, Exclusive OR gates).
2. Enable (G) should be (low, high) to pass the input to the output in the flip-flop in Figure 3–14.

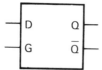

FIGURE 3–14

3. Enable (G) should be (low, high) to pass the input to the output in the flip-flop in Figure 3–15.

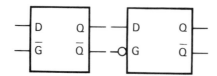

FIGURE 3–15

4. Using the timing diagram in Figure 3–16, draw the output waveform for the flip-flop in Figure 3–15.

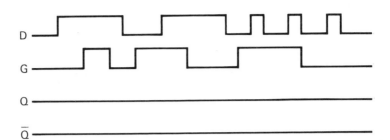

STOP   Do Experiment 3–2

FIGURE 3–16

Other D Latches

Some D latches transfer data only during the transition of the clock. That is, they do not transmit data for the length of the clock pulse but only during the high-to-low or low-to-high transition. These D latches are used in circuits where timing of data flow must be matched to other parts of a complete circuit. The output of these circuits changes state at the clock transition and remains in that state until the next transition. That is, if data is transferred on a high-to-low transition, the output changes states only on high-to-low transitions.

JK FLIP-FLOPS

JK flip-flops are the most versatile flip-flops because they can perform all the functions of RS and D-type flip-flops, plus much more. This versatility adds to their complexity as well as their cost.

A JK flip-flop is actually two flip-flops in one, the two feeding each other with proper inputs. The logic diagram for a *master–slave JK flip-flop* is shown in Figure 3–17. The master latch is at the input, and the slave latch is at the output.

The JK inputs set the master latch, which, in turn, feeds the slave latch. Both latches are controlled by a clock pulse. Although there are two places to store the data, the flip-flop is set or reset only by the slave latch. The master latch controls the state of the slave latch.

Gates 3 and 4 make up the master latch, which is controlled by input gates 1 (J) and 2 (K). Gates 7 and 8 make up the slave latch, which is controlled by gates 5 and 6. The clock (CLK) input controls the operation of the flip-flop, while the J and K inputs determine how to control the master latch. The inverter ensures that the clock is complementary for the master and slave latches.

Two other inputs that control the JK flip-flop are the set (S) and clear (C) inputs. These inputs override all other circuitry and are used to preset the flip-flop prior to any other operation. Note that the S input is low and the C input is high to set the JK flip-flop and the slave latch. The output Q of a set JK flip-flop is high. To reset, C is low and S is high. Both S and C should be high for other operations. Both S and C should not be low, or an unstable state will result similar to the one we discussed with RS flip-flops.

To analyze the operation of the JK flip-flop, we convert its func-

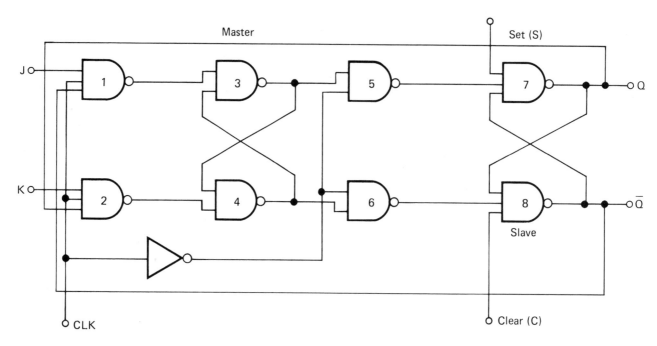

FIGURE 3–17

Master–Slave JK Flip-Flop

Condition		Set	Clear	Clock	Input		Output	
		S	C	CLK	J	K	Q	$\overline{Q}$
1	Asynchronous	0	1	X	X	X	1	0
2		1	0	X	X	X	0	1
3		0	0	X	X	X	Not allowed	
4	Synchronous	1	1	⎍↴	0	0	Q_o	$\overline{Q}_o$
5		1	1	⎍↴	1	0	1	0
6		1	1	⎍↴	0	1	0	1
7		1	1	⎍↴	1	1	Toggles*	

FIGURE 3–18

JK Flip-Flop Truth Table with Set and Clear

X = don't care

Q_o and $\overline{Q}_o$ = remains in previous state

*Toggles = changes to the other state

tions into a truth table, as shown in Figure 3–18, and focus on each possible condition of the inputs. *Asynchronous* operations conform to conditions 1, 2, and 3 of the truth table, while *synchronous* operations pertain to conditions 4 through 7.

Asynchronous Operations

Condition 1. The set (S) and clear (C) functions override all other functions. That is, they "don't care" what the CLK, J, or K state is. If S is low and C is high, the output Q will be set to one and $\overline{Q}$ will be zero.

Condition 2.   Again, CLK, J, and K are not part of the operation. The output Q will be cleared with the low pulse at C. Q will go low, and $\overline{Q}$ will be high. Note that the JK input is usually cleared or set immediately after power-on to assure its beginning state.

Condition 3.   If both S and C go low at exactly the same time, the output of Q and $\overline{Q}$ is unstable. It will not stay in any particular state when the inputs are returned to high. This condition must be avoided.

Synchronous Operations

Condition 4.   When both S and C are high, they are inactive and do not affect the circuit operation. J, K, and CLK now control the outputs. The output will not change states with J and K low during the complete CLK pulse. It will stay at whatever state it was in before the CLK pulse.

Condition 5.   S and C are inactive. When the CLK pulse arrives, J is high and K is low; so the output Q goes high. The J pulse can occur at any time during the CLK pulse, but K must be low for the whole CLK pulse. The actual change in output does not occur, however, until the trailing edge of the CLK pulse. See Figure 3–19.

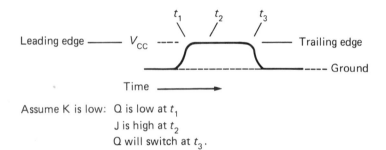

FIGURE 3–19

Clock Pulse for JK Flip-Flop

Condition 6.   S and C are again inactive. K is high and J is low; so the output Q goes low. J must be low for the whole CLK pulse, but K can be high at any time during the pulse. As in the previous condition, the output Q will make its transition on the trailing edge of the CLK pulse.

Condition 7.   S and C are inactive once again. J and K are high. The output will change states, regardless of whether it was high or low before the CLK pulse. The output will *toggle* (change) to the other state at the trailing edge of the CLK pulse. J and K do not have to be high at the same time, as long as they both are high at some time during the CLK pulse. Note here that, if J went high and then K went high and then J went high again, the output would still toggle. How many times the pulse on J or K occurs does not matter. Remember, no change occurs to the output without a CLK pulse unless $\overline{S}$ and $\overline{C}$ are used.

■ **EXAMPLE**

Figure 3–20 shows an example of a JK flip-flop timing diagram. We want to analyze the timing pulses in relation to the conditions of the truth table just discussed.

Solution

First, we mark the trailing edges of the CLK pulse, since the synchronous operations change the output state only at this point in time. We use an

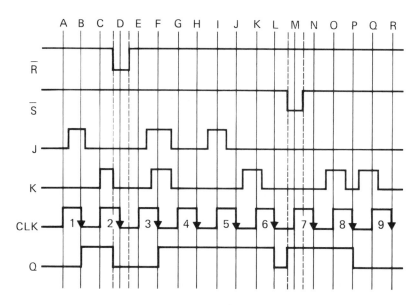

FIGURE 3–20

JK Flip-Flop Timing Diagram

arrow to emphasize each point and draw windows around the high part of the CLK pulse to help determine what happened during the pulse.

Next, we emphasize the asynchronous operations—that is, the $\overline{R}$ and $\overline{S}$ functions. We use a dashed line or a different color pencil.

Now, we analyze the timing pulses and draw the Q output. We assume Q is low to start. Seeing no asynchronous signals ($\overline{R}$ or $\overline{S}$) from point A to point B, we know that Q must stay in its previous state at least until the end of CLK pulse 1. We draw Q to point B as shown.

By looking at what happened during CLK pulse 1, we see that J went high and K stayed low for the whole CLK pulse (condition 5 of the truth table). Q must go high. We draw it. It must stay high until an asychronous pulse is seen, or until the end of CLK pulse 2.

We see that $\overline{R}$ goes low before the end of CLK pulse 2; so we draw Q high to that point, where it must go low to reset. As long as reset is low, the CLK pulses are ignored and no change is made at the end of CLK pulse 2.

So, we must now move on to the next asynchronous pulse, or the end of CLK pulse 3. No $\overline{S}$ or $\overline{R}$ pulses are found; so we draw Q to the end of CLK pulse 3 in its present low state. By looking back at what happened during CLK pulse 3, we see that both J and K were high at some time during the pulse (condition 7 of the truth table). The output Q must toggle. Since it is low now, it must go high at point F.

We see no $\overline{R}$ or $\overline{S}$ pulses from point F to point H; so we draw Q to point H in its previous state (high). There are no pulses on the J or K inputs during CLK pulse 4 (condition 4). There is no change to the output Q. We continue on to the end of CLK pulse 5 since no pulses are found on $\overline{R}$ or $\overline{S}$.

Although J went high during CLK pulse 5 (condition 5), since Q is already high no change will occur at this point. We go to the end of CLK pulse 6 since there are still no pulses on $\overline{R}$ or $\overline{S}$. K went high during CLK pulse 6, which means (according to the truth table) that Q goes low. We draw it.

We see that $\overline{S}$ has a low before we get to the end of CLK pulse 7. Therefore, we must show a set immediately when we reach that point and maintain it to the end of CLK pulse 7. Since, during CLK pulse 7, no signals were received at inputs J and K, Q stays in its previous state.

We now move on to the end of CLK pulse 8. A signal was seen on input K; so the output Q must go low (condition 6). At the end of CLK pulse 9, we see that the only signal was a high on input K. Q must go low. Since it is already low, no change occurs here. Our analysis is finished. ■

■ EXAMPLE

Analyze Figure 3–21 in the same way as you did for Figure 3–20. Follow it through until you are familiar with how the truth table relates to the actual timing diagram.

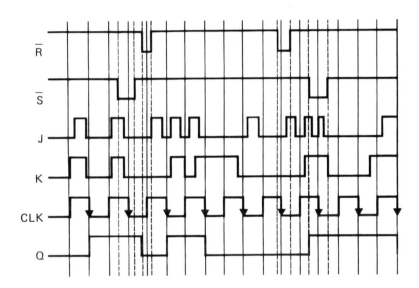

FIGURE 3–21
JK Flip-Flop Timing Diagram

STOP   Do Experiment 3–3

SELF-TEST EXERCISE 3–3

1. The JK flip-flop is a two-in-one flip-flop. True or False?
2. The slave latch determines the state of the flip-flop. True of False?
3. The master and slave latches are controlled by the same clock pulses—master, by leading edge and slave, by trailing edge. True or False?
4. When $\overline{Q} = Q$, the flip-flop is in a (set, reset, toggling, unacceptable) state.
5. When J and K = 1, the flip-flop will _____ after the clock pulse.
6. When J = 1 and K = 0, $\overline{Q}$ becomes (low, high) after the clock pulse.
7. When J = K = 0, the flip-flop (changes, does not change) states after the clock pulse.
8. S and C cannot do anything during the clock pulses. True or False?
9. When J and K = 1, $\overline{S}$ = 0, and $\overline{C}$ = 1, the flip-flop Q line will go (low, high) (after the clock pulse, immediately).

SUMMARY

A basic sequential logic memory element is used for temporary storage of data. It "remembers" a signal after the input is removed. It is often used in computer circuits to hold data until the proper timing signal releases the data for use. Temporary storage also means the signal is lost if power is lost.

RS and $\overline{RS}$ flip-flops are latches that are very useful where several signals may appear on a line but only the first signal is to be used. The RS latch responds to the first signal but is immune to any others until the opposing input is signaled. A debounce circuit operates in this manner.

D-type bistable flip-flops are used where many signals are applied to a line but only some of the signals are to be used. The enable input is used to allow the desired signals to pass through to the output, where they can be input to another device.

A simple storage register can hold only as many bits as it has flip-flops. A 4-bit register has four flip-flops and can temporarily store four bits of data at a time.

JK flip-flops have both synchronous and asynchronous operations. Synchronous operations must operate with a clock signal. Asynchronous operations occur at any time. The S and C inputs of the JK flip-flop are asynchronous. The J and K inputs must operate with the clock signal to produce an output.

A simple divide-by-two circuit can be built by tying the S, C, J, and K inputs of a JK flip-flop to V_{cc} and applying to the clock input the frequency to be divided.

The leading edge of a clock pulse is the low-to-high transition. The trailing edge of a clock pulse is the high-to-low transition. It is very important for you to know the difference when timing of data flow is a consideration in a circuit.

CHAPTER 3
REVIEW EXERCISES

1. The Q output of a flip-flop is high. What state is the flip-flop in?

   a. set
   b. reset

2. The *complement* output of a flip-flop is low. What binary state is stored in the flip-flop?

   a. binary 0
   b. binary 1

3. A storage register made up of six D-type flip-flops is storing a binary number. The flip-flop states are as follows: A = reset, B = set, C = reset, D = reset, E = set, and F = set. The A flip-flop is the LSB. What is the decimal equivalent of the register content?

   a. 16
   b. 19
   c. 36
   d. 50

4. How many JK flip-flops are needed to generate a 125 kHz square wave from a 1 MHz square wave?

a. 1
b. 2
c. 3
d. 4

5. Given the input waveforms shown, draw the Q output of the $\overline{RS}$ flip-flop in Figure 3–22.

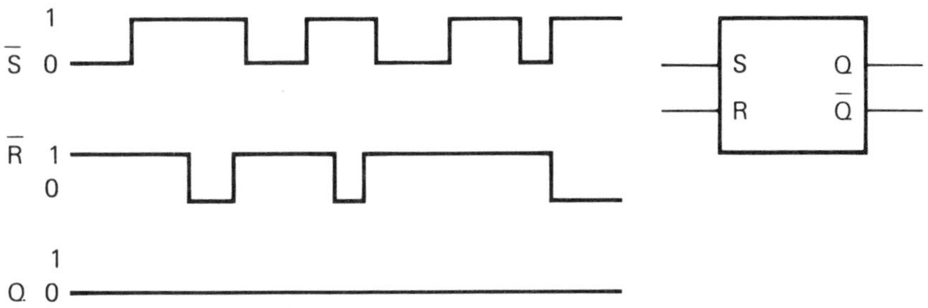

FIGURE 3–22

6. Given the input waveforms shown, draw the Q output of the RS flip-flop in Figure 3–23.

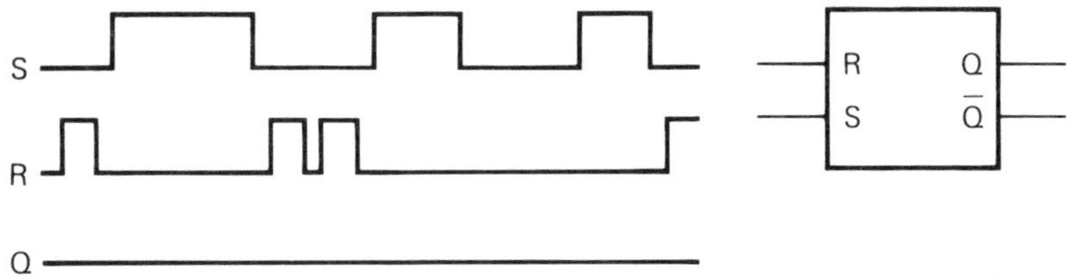

FIGURE 3–23

7. Will the flip-flop in Figure 3–24 operate properly with these inputs for an RS latch? If not, where is the error?

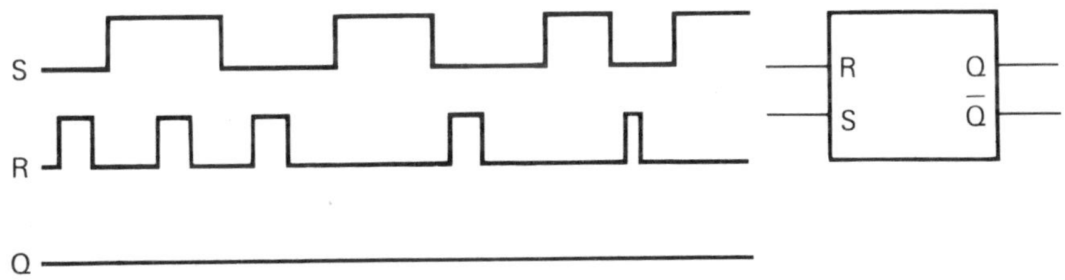

FIGURE 3–24

8. Given the input waveforms shown, draw the normal output of the flip-flop in Figure 3–25.

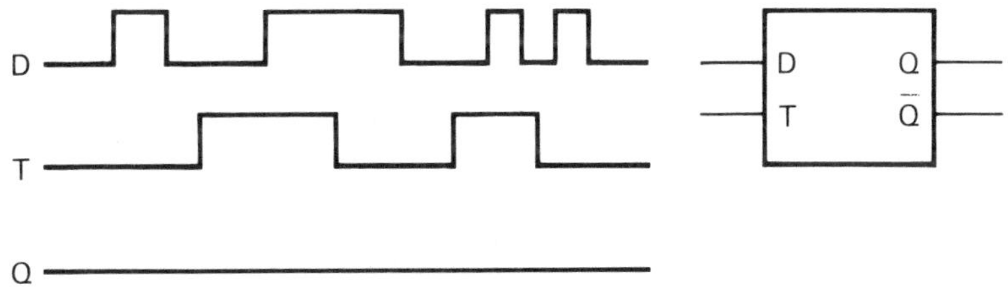

FIGURE 3–25

9. Given the input waveforms shown, draw the normal output of the flip-flop in Figure 3–26.

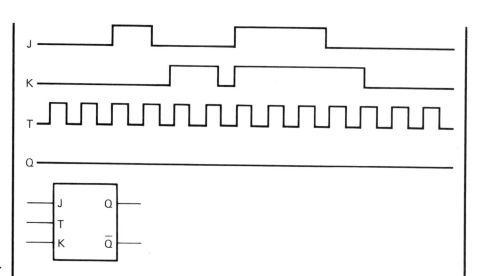

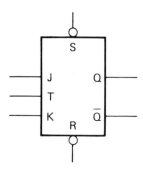

FIGURE 3–26

10. Check all of the following ways in which a JK flip-flop can be set. (See Figure 3–27.)

  a. Ground the C input.
  b. Ground the S input.
  c. Set J to 0, K to 1, and apply a clock pulse.
  d. Set J to 1, K to 0, and apply a clock pulse.
  e. Toggle the T input with J = K = 1, if it is low to start.

FIGURE 3–27

11. How many flip-flops will be needed to store the binary equivalent of the decimal number 114?

  a. 3
  b. 7
  c. 12
  d. 57

12. The trailing edge of a logic signal generally refers to its transition from:

  a. 0 to 1
  b. 1 to 0
  c. either 0 to 1 or 1 to 0

13. For a 7476 JK flip-flop to work properly in the synchronous mode, the S and C input states must be:

a. S = 0, C = 0
b. S = 0, C = 1
c. S = 1, C = 0
d. S = 1, C = 1

14. When the T input to a D-type flip-flop is high and the D input is a logic signal X, the complement output is:

   a. binary 0
   b. binary 1
   c. $\overline{X}$
   d. X

15. The basic application of a D-type flip-flop as given in this chapter is:

   a. switch contact debouncing
   b. storage
   c. frequency division
   d. counting

16. Which type of flip-flop does not have an ambiguous state?

   a. RS
   b. D
   c. JK

17. In a JK flip-flop, exactly when does the output Q change states in the toggle mode of operation?

EXPERIMENT 3–1

Memory Elements and RS Flip-Flops

PURPOSE

Part I of this experiment is designed to show the operation of a basic memory element; Part II will deal with RS flip-flops. You will complete a truth table for each RS latch. The RS latches will be built from basic gates so that you will not only gain experience in breadboarding circuits but also clearly see the operation of the circuit.

EQUIPMENT

—1 logic probe with pulse indicator (Equip$_1$)
—1 digital experimenter (Equip$_2$)
—1 SN7432 IC OR gate (U_1)
—1 SN7402 IC NOR gate (U_2)
—1 SN7400 IC NAND gate (U_3)

PROCEDURE: PART I

Step 1 Install U_1 in the breadboard on the experimenter.

Step 2 Wire the circuit shown in Figure 3E1–1 before you turn on the experimenter. Pin numbers are given. Pin 14 is V_{CC} +5 V. Pin 7 is ground.

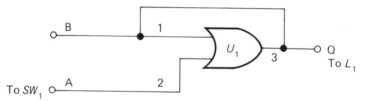

FIGURE 3E1–1

Step 3 Data switch SW_1 should be low. Wire point B to data switch SW_2, with SW_2 in the low position.

Step 4 Turn on power. Logic indicator L_1 should be off.

Step 5 Remove the wire to SW_2.
NOTE: L_1 is still off because Q is low.

Step 6 Switch SW_1 to high. L_1 will light.

Step 7 Move SW_1 to low. L_1 is still on; it remembers the input signal.

Step 8 Switch SW_1 back and forth several times; no change will be seen.
NOTE: The only way the output can be taken low again is to repeat this procedure.

Check: Can you think of any application for such a circuit?

NOTE: There are very few applications for a circuit that is so difficult to reset. The RS latch in Part II is much more versatile.

PROCEDURE: PART II

Step 1 Wire the RS latch circuit as shown in Figure 3E1–2. Do not forget to connect +5 V and ground to IC pins 14 and 7, respectively.

NOTE: The S (set) and R (reset) inputs come from logic switches A and B. The latch outputs Q and $\overline{Q}$ can be seen with a logic probe or wired to logic indicators L_1 and L_2.

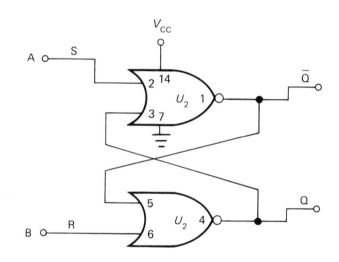

FIGURE 3E1–2

Step 2 Apply power. Observe the state of the inputs and outputs. They can be checked with a logic probe.

ACTIVITY

Record the results:

$$\text{set} = A = \underline{\hspace{2cm}}$$
$$\text{reset} = B = \underline{\hspace{2cm}}$$
$$Q = L_1 = \underline{\hspace{2cm}}$$
$$\overline{Q} = L_2 = \underline{\hspace{2cm}}$$

Step 3 Apply different logic levels to the set and reset inputs.

NOTE: Keep in mind that the logic switch sends a positive pulse to the input and then settles back to a low input, which the logic probe should confirm if you measure the inputs while you move the switches.

ACTIVITY

Complete the truth table in Figure 3E1–3.

S (A)	R (B)	Q (L_1)	$\overline{Q}$ (L_2)	Explanation of state
0	0			
0	1			
1	0			
1	1			

FIGURE 3E1–3

Step 4 Randomly go through different inputs and observe the outputs. This step will help you understand the latch.

Step 5 Construct the $\overline{RS}$ latch shown in Figure 3E1–4 using NAND gates.

NOTE: The $\overline{S}$ (set) and $\overline{R}$ (reset) inputs come from logic switches $\overline{A}$ and $\overline{B}$. The inputs are connected to $\overline{A}$ and $\overline{B}$ because this circuit operates on an active low or negative pulse. The latch outputs Q and $\overline{Q}$ can be connected to logic indicators L_1 and L_2 or checked with a logic probe.

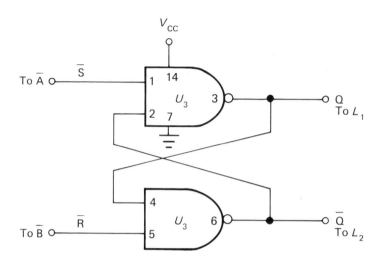

FIGURE 3E1–4

Step 6 Apply power. Observe the state of the inputs and outputs. They can be checked with a logic probe.

ACTIVITY

Record the results:

$$set = A = \underline{\hspace{2cm}}$$
$$reset = B = \underline{\hspace{2cm}}$$
$$Q = L_1 = \underline{\hspace{2cm}}$$
$$\overline{Q} = L_2 = \underline{\hspace{2cm}}$$

Step 7 Apply different logic levels to the $\overline{set}$ and $\overline{reset}$ inputs.

ACTIVITY

Complete the truth table in Figure 3E1–5.

NOTE: Remember, when you are connected to $\overline{A}$ and $\overline{B}$ logic switches, the inputs are normally high until you switch them low for a negative input pulse to the flip-flop.

FIGURE 3E1–5

$\overline{S}$ (A)	$\overline{R}$ (B)	Q (L_1)	$\overline{Q}$ (L_2)	Explanation of state
0	0			
0	1			
1	0			
1	1			

Step 8 Repeat this experiment until you understand the operation of the RS and $\overline{\text{RS}}$ flip-flops.

NOTE: You will see from the truth table that a high sets an RS latch while a low sets an $\overline{\text{RS}}$ latch.

EXPERIMENT 3–2

D-Type Bistable Flip-Flops and Storage Registers

PURPOSE

This experiment is designed to show the operation of a D-type bistable flip-flop. You will build it from individual gates in integrated circuit form. You will then build a storage register from a D-type latch IC. Truth tables will be completed for each latch.

EQUIPMENT

—1 logic probe (Equip$_1$)
—1 digital experimenter (Equip$_2$)
—1 7400 IC NAND gate (U_1)
—1 7402 IC NOR gate (U_2)
—1 7475 IC bistable latch (U_3)

PROCEDURE

Step 1 Wire the circuit shown in Figure 3E2–1 using the 7400 IC. Data switches SW_1 and SW_2 supply inputs at D and G. Q and $\overline{Q}$ are connected to logic indicators L_1 and L_2. A logic probe can be used to monitor the inputs and outputs if desired.

NOTE: This circuit can be purchased in IC form, but, to help you understand its operation, it will be wired from individual gates in this experiment.

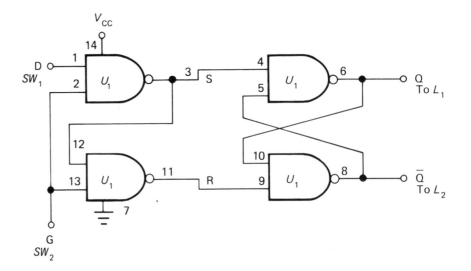

FIGURE 3E2–1

Step 2 Set the input data switch SW_1 (D) to zero and the enable switch SW_2 (G) to one.

ACTIVITY

Record the output in the truth table in Figure 3E2–2. Then apply different logic levels to the inputs and record the results in the table.

D	G	Q (L_1)	$\overline{Q}$ (L_2)
0	1		
0	0		
1	1		
1	0		

FIGURE 3E2–2

Step 3 Apply a 1 Hz clock to the D input for more dynamic testing. Leave the G input at SW_2.

ACTIVITY

Record the output at Q with G (SW_2) at one.

Step 4 Move G (SW_2) to zero.

ACTIVITY

Record the output at Q with G (SW_2) at zero.

Step 5 Construct the circuit shown in Figure 3E2–3.

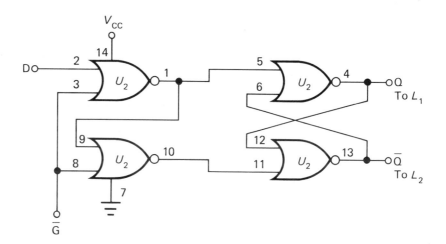

FIGURE 3E2–3

ACTIVITY

Set switches D and $\overline{G}$ as shown in the truth table in Figure 3E2–4 and record the outputs in the table. Then, try various switch positions and note how the flip-flop operates.

D	$\overline{G}$	Q	$\overline{Q}$
1	0		
1	1		
0	1		
0	0		

FIGURE 3E2–4

Step 6 Apply a 1 Hz input to the D input.

ACTIVITY

Record the output at Q with $\overline{G}$ set at one.

Check: What is the difference between the two circuits you have just connected?

One circuit works on a high enable; one, on a low enable.

Step 7 Wire the circuit shown in Figure 3E2–5. The 7475 IC contains four TTL bistable flip-flops or latches. It is called a 4-bit bistable latch because it has four separate Q lines out. Four data switches will be used as the input to the register. Monitor the output by using logic indicators as shown. This register works on a high enable. Pins 13 and 4 are the control lines for the four latches. They are to be connected to logic switch A. (A logic switch is a momentary push button that gives a low or high pulse out). Logic switch A is normally low but sends a high pulse when the push button is depressed.

NOTE: Do not forget to wire up the V_{CC} and ground pins. Pin 12 is ground, and pin 5 is V_{CC}.

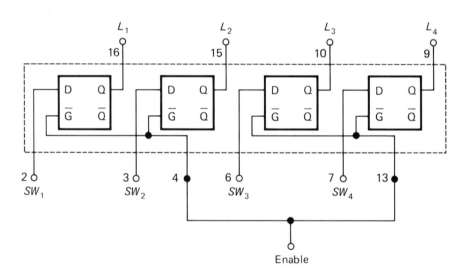

FIGURE 3E2–5

Step 8 Apply power to the circuit.

Step 9 Set the four data switches to 0000. Logic switch A is at zero.

Step 10 Depress logic switch A.

NOTE: This should clear all flip-flops by sending a high pulse to the load input; 0000 data is loaded into the registers. All logic indicators should be off.

Step 11 Set the four data switches to 1001. Depress logic switch A and release while you monitor the output.

NOTE: 1001 should be displayed at the logic indicators since this 4-bit word has now been loaded.

Step 12 Set the four data switches to 1111. Depress logic switch A and release while you monitor the output.

NOTE: Depressing A should have loaded the register with the 4-bit word 1111. The logic indicators should all be on.

Step 13 Try various other 4-bit input combinations. The output will not change until the enable line is pulsed high.

Check: The input is passed on to the output only when the enable goes _____.

high

Check: The output depends on the input and the state of the _____ line.

enable

| EXPERIMENT 3–3 | # JK Flip-Flops with Set and Clear |

PURPOSE

This experiment is designed to familiarize you with the operation of the JK flip-flop. You will complete truth tables and draw timing diagrams to help you understand how the circuit works.

EQUIPMENT

—1 logic probe (Equip₁)
—1 digital experimenter (Equip₂)
—1 dual-trace oscilloscope (Equip₃)
—1 SN7476 IC JK flip-flop (U_1)

PROCEDURE

Step 1 For the circuit shown in Figure 3E3–1, connect +5 V to pin 5 and ground to pin 13.

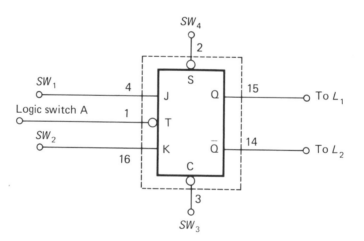

FIGURE 3E3–1

Step 2 Then connect the rest of the circuit as shown. Use the data switches for J, K, S, and C. Use logic switch A for the clock (T) input. Connect logic indicators to the outputs Q and Q̄ or use a logic probe to monitor them.

NOTE: There are two JK flip-flops in an SN7476 IC, as a data sheet for the IC will show. We are using only one of the flip-flops here.

Step 3 Set J = K = 1 with SW_1 and SW_2 to check the S and C operation.

ACTIVITY

Apply the signal for S and C as shown in the truth table in Figure 3E3–2 and record the results. Set J and K to different levels while you are doing this. Notice what happens when S and C are both low. Remember this condition as an unacceptable state and avoid it. To inhibit or make the set and clear state inactive, hold both S and C high.

NOTE: The demonstration of the asynchronous operation of this JK flop is now complete. You have learned that asynchronous means that the action of the flip-flop is not dominated by a clock pulse. It can happen at any time when proper voltage levels are applied to S and C. S and C are the asynchronous operation in this case. In the following steps, you will verify the operation of the JK flip-flop using the J and K inputs.

Input					Output	
J	K	S	C	G, CLK	Q	$\overline{Q}$
		1	0	0		
		0	1	0		
		1	1	0		
		0	0	0		

FIGURE 3E3–2

Step 4 Set S and C to one with switches SW_3 and SW_4.

ACTIVITY

Apply the logic levels to J and K as shown in the truth table in Figure 3E3–3. Record the output states before the clock pulse (Q_t) and after the clock pulse (Q_{t+1}) using logic switch A for the CLK input. After you complete the truth table, repeat each step, applying the clock pulse a number of times and noting the changes in the output.

J	K	Q_t	Q_{t+1}
0	0		
0	1		
1	0		
1	1		

FIGURE 3E3–3

Step 5 Remove the logic switch A connection to the CLK input and connect CLK to a 1 Hz clock.

ACTIVITY

Check the output by setting J and K to the different logic inputs shown in Figure 3E3–3.

NOTE: You have just demonstrated the synchronous operation of a JK flip-flop and should understand that in this mode the output is controlled by a clock pulse. Synchronous operation is more critical than asynchronous but requires fewer and simpler circuits. The output has a constant time interval between pulses out. As you will discover in the following steps, the JK flip-flop is often used as a divider to divide the clock pulse.

Step 6 Construct the circuit shown in Figure 3E3–4. The circuit is driven by a 1 kHz clock signal. The $\overline{A}$ logic switch is used to control the J and K inputs. The $\overline{B}$ logic switch controls the clear function.

NOTE: Although any input pins left open are essentially at logic one, it is always best to tie them to V_{CC} to prevent them from picking up any stray signals or noise. The SN7476 is especially sensitive to noise in this manner. You can test for this problem by trying to make the circuit operate properly with these inputs open.

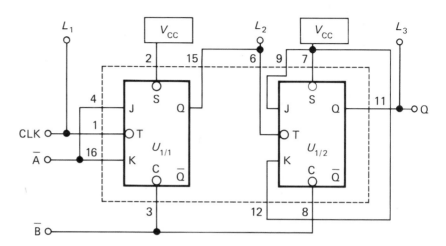

FIGURE 3E3–4

ACTIVITY

Compare the waveforms at pins 1, 6, and 11. Draw a timing diagram. Ignore L_1, L_2, and L_3 in this activity. Since the 1 kHz signal is too fast for you to monitor on the logic indicators, use the dual-trace oscilloscope.

NOTE: Your timing diagram should look like the one in Figure 3E3–5. You can see that each input is divided in half by the JK toggle action. The clock input to JK $U_{1/1}$ is divided in half, thereby producing the output seen at pin 6. This output is fed to the JK $U_{1/2}$ input, which is divided in half by this flip-flop, producing the output at pin 11.

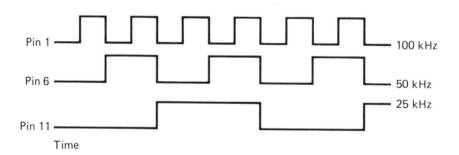

FIGURE 3E3–5

Step 7 Change the clock pulse to 1 Hz so that you can monitor the following with the logic indicators.

Step 8 Depress logic switch $\overline{A}$, which puts a low pulse to the JK inputs.

NOTE: The low input inhibits the first flip-flop from pulsing, and, thus, the second flip-flop does not pulse. The output is locked on to the state set by the clock pulse prior to J and K going low. This result agrees with the truth table.

Step 9 Depress logic switch $\overline{B}$ and check the Q output during operation. $\overline{B}$ is normally high.

NOTE: Depressing the switch applies a low to both flip-flops. The output is cleared. The clock, J, and K have no affect.

Counter Circuits

After studying this chapter, you will be able to:

1. Describe the operation of a binary counter.
2. Calculate the maximum decimal count for an N-bit binary counter (N is the number of bits).
3. Describe the input/output characteristics of binary up, down, and up/down counters.
4. Describe the operation of a ripple counter. Explain what cascading flip-flops are.
5. Describe the operation of a decade counter.
6. Calculate the maximum decimal count for N number of stages in a decade counter.
7. Calculate the output frequency of a binary, decade, or combination counter circuit, given the input frequency.
8. Determine the number of pulses needed for a specific binary output of a binary or decade counter.
9. Determine the number of decade counters needed for a particular decimal output.
10. Give examples of and describe the functions of control pins that are commonly found on IC counters.
11. Determine the number of states and maximum decimal count of a modulo N counter.

INTRODUCTION

Counter circuits are used when mathematical as well as timing functions are required. Counters can be made to communicate in binary or digital format. This chapter provides basic information on digital counter circuits.

First, we must define what a counter is. A *counter*, also called a *divider*, is a sequential logic circuit made up of flip-flops. The important word here is *sequential*, which means one after the other in an orderly pattern—that is, consecutive or serial. A counter is used to count the number of pulses applied to it. These pulses can represent people, or products, or the number of specific events that occur over a period of time. Pulses are applied to a counter input and cause the flip-flops to change state in such a way that the binary number stored in the flip-flops is equal to the number of input pulses.

There are many different types of counters. The most important ones are the binary and the decade counters. Counters can operate in synchronous or asynchronous modes.

ASYNCHRONOUS COUNTERS

Binary Ripple Up Counter

A *binary counter*, also referred to as a *binary scaler*, uses the pure binary code. A binary *ripple counter* is formed by cascading JK flip-flops. *Cascading* occurs when two or more circuits are connected in series so that the output from one circuit provides the input to the next, as shown in Figure 4–1.

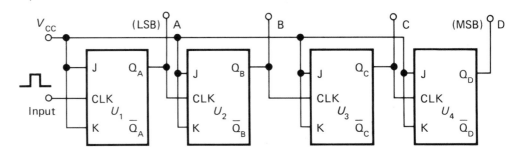

FIGURE 4–1

4-Bit Binary Ripple Up Counter

The Q output of each flip-flop in Figure 4–1 is connected to the clock (CLK) input of the next flip-flop to form a *binary ripple up counter*. Since all the JK inputs are held high, the output of each flip-flop toggles when its clock input sees a high-to-low transition. The input pulses to be counted are applied to the clock input of the first flip-flop, U_1. The counter operates as follows.

When the first clock pulse trailing edge is applied to U_1, the Q_A output sets. To read the number stored in the flip-flop, note that U_1 is the *least significant bit* (LSB) and D is the *most significant bit* (MSB) in the binary word output. Therefore, after the first pulse, we have a binary word of 0001_2 (DCBA), which is equal to a decimal 1. See Figure 4–2.

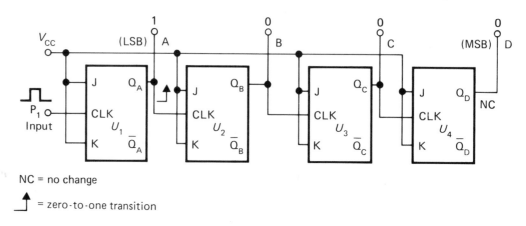

NC = no change

= zero-to-one transition

FIGURE 4–2

Ripple Up Counter Clock Pulse 1

When the second input pulse is applied to U_1, Q_A toggles to the low state and causes a one-to-zero change at the clock input of U_2. The Q_B output then toggles from zero to one. Therefore, after two pulses, we have a binary word of 0010_2 (DCBA), which is equal to a decimal 2. See Figure 4–3.

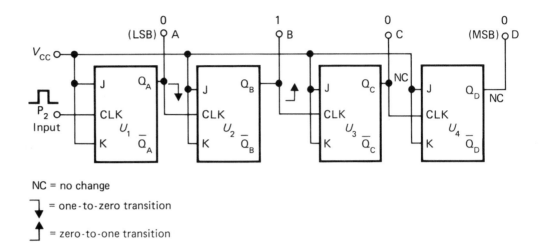

NC = no change

⬐ = one-to-zero transition

⬏ = zero-to-one transition

FIGURE 4–3

Ripple Up Counter Clock Pulse 2

The third input pulse causes Q_A to toggle from zero to one. No change occurs at Q_B. Therefore, the output of the counter is now a binary word of 0011_2 (DCBA), or a decimal 3. See Figure 4–4.

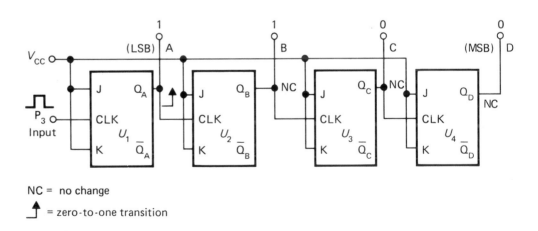

NC = no change

⬏ = zero-to-one transition

FIGURE 4–4

Ripple Up Counter Clock Pulse 3

The fourth input pulse causes Q_A to toggle from one to zero, which clocks U_2 output Q_B from one to zero. The Q_B transition toggles the Q_C output. So, the output is the binary word 0100_2 (DCBA), or decimal 4. See Figure 4–5.

In summary, each flip-flop's high-to-low transition toggles the next flip-flop in line. The input thereby "ripples" through the counter and

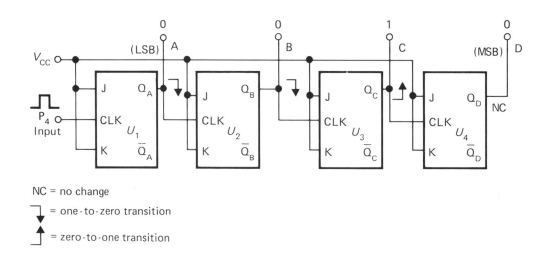

FIGURE 4–5

**Ripple Up Counter Clock
Pulse 4**

increases the number stored in the counter by one until the maximum count of 1111_2 (DCBA), or decimal 15, is reached. An important change occurs when all flip-flops are at one. That is, when the counter is at 1111_2, one more pulse changes Q_A to zero, which changes Q_B to zero, which changes Q_C to zero, which changes Q_D to zero. See Figure 4–6.

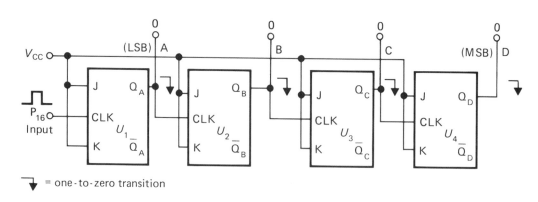

FIGURE 4–6

**Ripple Up Counter Clock
Pulse 16**

The count sequence at the outputs A, B, C, and D is summarized in the truth table in Figure 4–7. The input and output waveforms for the binary ripple up counter are shown in the timing diagram in Figure 4–8. Notice, from the truth table, that the binary output for this counter goes from 0000 to 1111. After decimal 15 (at decimal 16), the output goes to 0000_2. This change *recycles* (restarts) the count. Note also that, although in Figure 4–8 the input pulses are evenly spaced or of constant frequency, they do not have to be. The counting operation is the same either way.

This counter is called a 4-bit counter because there are four output bits—A, B, C, and D. The maximum number of states is calculated by

Count	D	C	B	A
0	0	0	0	0
1	0	0	0	1
2	0	0	1	0
3	0	0	1	1
4	0	1	0	0
5	0	1	0	1
6	0	1	1	0
7	0	1	1	1
8	1	0	0	0
9	1	0	0	1
10	1	0	1	0
11	1	0	1	1
12	1	1	0	0
13	1	1	0	1
14	1	1	1	0
15	1	1	1	1

Recycle

FIGURE 4–7

Count Sequence for a 4-Bit Binary Ripple Up Counter

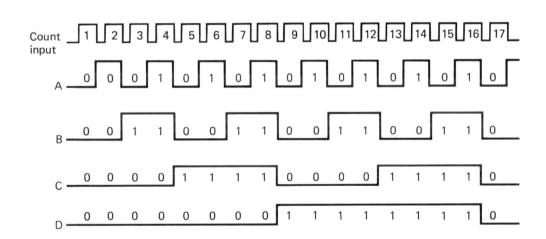

FIGURE 4–8

Timing Diagram for a 4-Bit Binary Ripple Up Counter

using 2^N where N is the number of bits. The maximum decimal output is $2^N - 1$. For example, a 5-bit counter counts 2^5, or 32, pulses. The maximum decimal number that it can represent is 31 (11111_2 or $2^5 - 1$).

SELF-TEST EXERCISE 4–1

1. A counter is also referred to as a divider. True or False?
2. A 6-bit counter counts up to decimal output _____. It has _____ states.
3. A counter is made up of flip-flops. True or False?
4. Any number of flip-flops can be _____ to form a counter.
5. Counters must have a constant frequency input to count with no error. True or False?

6. A counter is a _____ logic circuit made up of flip-flops.
7. Counters only operate in a synchronous mode. True or False?
8. The output of a binary up counter after _____ input pulses will be 0110_2. (Assume that it was cleared to start the count.)

Binary Ripple Down Counter

In a *binary ripple down counter*, each input pulse causes the binary number stored in the counter to *decrease* by one. That is, the input pulses are used to decrement the count. A 4-bit binary ripple down counter is shown in Figure 4–9.

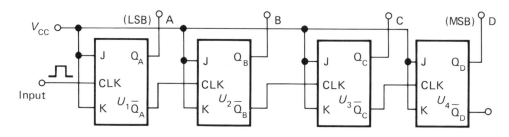

FIGURE 4–9

4-Bit Binary Ripple Down Counter

The difference between the circuits of an up and a down counter is that the toggle inputs of each stage of a down counter are fed from the complement output of the previous stage. Thus, the count sequence is the exact reverse of that of the up counter. See Figure 4–10. Notice that the count sequence of the down counter is represented in the same way as it was for the up counter—that is, by the Q outputs of each flip-flop. The input and output waveforms for the binary ripple down counter are shown in the timing diagram in Figure 4–11.

Input pulse	Output count	D	C	B	A
1	15	1	1	1	1
2	14	1	1	1	0
3	13	1	1	0	1
4	12	1	1	0	0
5	11	1	0	1	1
6	10	1	0	1	0
7	9	1	0	0	1
8	8	1	0	0	0
9	7	0	1	1	1
10	6	0	1	1	0
11	5	0	1	0	1
12	4	0	1	0	0
13	3	0	0	1	1
14	2	0	0	1	0
15	1	0	0	0	1
16	0	0	0	0	0

Recycle

FIGURE 4–10

Count Sequence for a 4-Bit Binary Ripple Down Counter

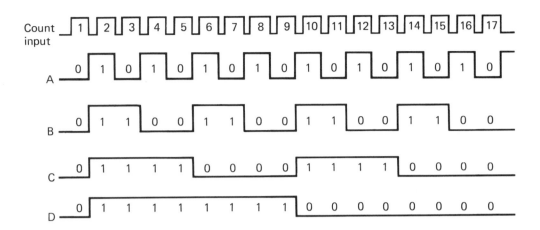

FIGURE 4–11

Timing Diagram for a 4-Bit
Binary Ripple Down Counter

Figure 4–12 shows one sequence of what happens in a binary down counter. If we assume an output of 0110_2 (DCBA), or decimal 6, the next input pulse will toggle the $\overline{Q}_A$ output to a low, which will toggle U_2. Q_B will then go low, and $\overline{Q}_B$ will go high. Remember, the outputs toggle only at the trailing edge, or the one-to-zero transition. Since U_2's $\overline{Q}_B$ output was low, it will toggle to a high. This transition will not toggle U_3, which means that U_4 will not toggle either. The output of the counter will be 0101_2 (DCBA), or decimal 5. In effect, the counter has counted down one pulse.

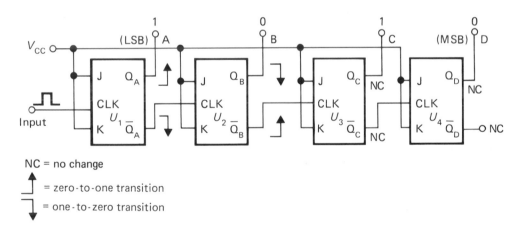

NC = no change

⌐ = zero-to-one transition

⌐ = one-to-zero transition

FIGURE 4–12

Ripple Down Counter
Clock Pulse

Although the operation of up and down counters has been described with separate JK flip-flops, it should be noted that they exist in IC form. An IC can be purchased that has up and down counting capability. A 4-bit counter is one IC.

**SELF-TEST
EXERCISE 4–2**

1. The input pulse (increments, decrements) the count in a down counter.
2. The (Q, $\overline{Q}$) output feeds the clock input of successive flip-flops in a down counter.

3. Counters can be purchased in integrated circuit form. True or False?
4. Every flip-flop triggers the next flip-flop in (synchronous, asynchronous) counters.
5. An 8-bit binary down counter can count from _____ to zero.
    a. 0
    b. 254
    c. 255
    d. 256
    e. 257

Binary Ripple Up/Down Counter

ICs are available that can count up as well as down depending on whether the input is applied to the up or the down input pin. A *binary ripple up/down counter* has a control pin that determines whether the count is up or down. See Figure 4–13. These ICs are used because of the complexity involved in building the circuits from individual JK flip-flops and using additional gates.

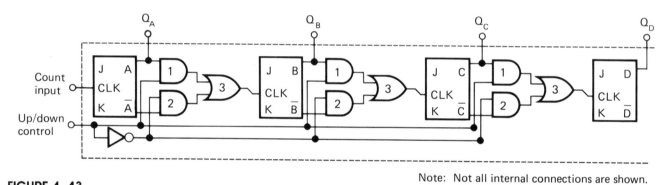

Note: Not all internal connections are shown.

FIGURE 4–13

Binary Ripple Up/Down Counter

COUNTER PROPAGATION DELAY

The ripple counters in the preceding section are *asynchronous* counters because the flip-flops are cascaded and one flip-flop triggers the next in sequence. That is, the count ripples through the counter, and the individual JK inputs are not controlled by a single simultaneous clock pulse.

Although ripple counters are simple to use, their main limitation is counting speed. The *propagation delay*—that is, the time delay between input pulse and resulting output pulse with no error—adds up as the number of flip-flops increases because the pulse has to ripple through all the stages before the count is settled. The maximum delay equals the propagation delay (in nanoseconds) for one flip-flop multiplied by the number of stages, assuming that we use the same kind of flip-flops. If each flip-flop, for example, has a 100 ns delay and there are four stages, the maximum delay is 400 ns. If pulses occurring at faster than 400 ns are fed, the counter will not count correctly. The input frequency limit (f) is found by the following formula:

$$f = \frac{1}{n \times t} \times 10^9$$

where n = number of flip-flops
       t = propagation delay (in nanoseconds) for one flip-flop

The following example shows how to figure the input frequency limit for a 4-bit ripple counter.

■ **EXAMPLE**

What is the maximum input counting frequency of four cascaded flip-flops when each flip-flop has a 25 ns delay?

Solution

$$f = \frac{1}{4 \times 25} \times 10^9 = 10 \text{ MHz}$$ ■

The counter in this example will not work properly if the input exceeds 10 MHz. Whenever possible, circuits are designed so that the maximum operating frequency of the counter far exceeds the input frequency.

The problem of propagation delay can be avoided by using flip-flops that are fast enough to work up to 500 MHz. But, since such fast circuits are so expensive and have high power consumption, synchronous counters are used.

All flip-flops in a *synchronous* counter are timed to a master timing signal or clock—that is, triggered simultaneously (at the same time) by the same clock pulse or the signal to be counted. Since all flip-flops change state at the same time, the propagation delay is equal to the delay of the slowest single flip-flop. Higher counting speeds are achieved because the delays are not cumulative.

SYNCHRONOUS COUNTERS

Synchronous Binary Up Counter

A *synchronous binary up counter* is shown in Figure 4–14. All the clock inputs are connected to a common line. The flip-flops are controlled by their J and K inputs. All the flip-flops are synchronized to the input to be counted. Like a ripple counter, a synchronous counter can have any number of flip-flops and corresponding outputs.

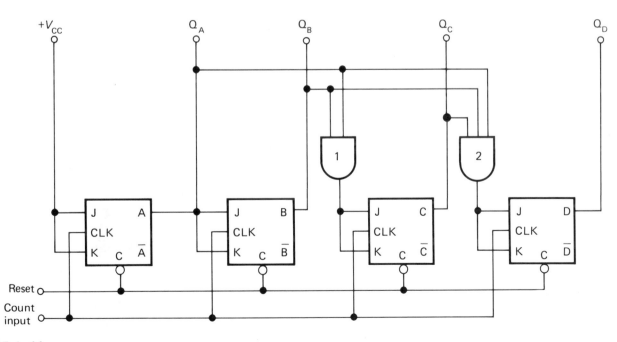

FIGURE 4–14

Synchronous Binary Up Counter

The counting sequence for a synchronous up counter is the same as it is for a ripple up counter. However, all the flip-flops change state at the same time. At no time is the output *ambiguous* (in an error state).

This is not the case in a ripple counter, where the outputs A, B, C, and D do not change to a new binary word at the same time and where there is a very short time during which the output is ambiguous. For example, if the output in a ripple counter is to change from binary 0111 to the next state 1000, it goes through the steps shown in Figure 4–15.

FIGURE 4–15

Ambiguous States in a Ripple Up Counter

D (MSB)	C	B	A (LSB)	
0	1	1	1	
0	1	1	0	
0	1	0	0	Ambiguous
0	0	0	0	
1	0	0	0	

It should be noted, however, that an ambiguous state is in the nanosecond range and would appear as a spike of only very short duration at the output. While the occurrence of ambiguous states could affect circuits that operate at a very fast speed, since all flip-flops in synchronous circuits change at the same time, there is no such problem.

SELF-TEST EXERCISE 4–3

1. Up/down counters count _____ depending on the control inputs or which input is used.
   a. up
   b. down
   c. up or down
2. In (synchronous, asynchronous) counters, all the flip-flops are controlled by a single clock pulse or the input pulse.
3. Propagation delay adds up in (synchronous, asynchronous) counters.
4. An 8-bit ripple counter has a propagation delay of _____ ns if each flip-flop has a delay of 30 ns.
5. The maximum counting frequency of the flip-flop in Question 4 is approximately _____ MHz.
6. The propagation delay of an 8-bit synchronous binary counter if each flip-flop has a delay of 25 ns is _____ ns.
7. The maximum operating frequency of the flip-flop in Question 6 is approximately _____ MHz.

 STOP Do Experiment 4–1

DECADE COUNTERS

Asynchronous Decade Up Counter

A *decade up counter*, also known as a BCD counter, decimal counter, or divide-by-two-and-divide-by-five counter, counts in the 8–4–2–1 binary code (BCD) up to 9. Four bits are needed to represent 0 through 9. The counter counts up for each pulse until the ninth count is reached; then, it recycles to 0000_2 and counts up again. It, therefore, has ten states.

The timing diagram for an asynchronous decade up counter is shown in Figure 4–16. The count sequence for the first nine input pulses (0–9) is the same as the one for a binary ripple counter. The operation that occurs at the tenth pulse is unique to the decade counter. Notice that, after the counter is at the ninth count (1001_2), if one more pulse is applied, the output goes to zero (0000_2).

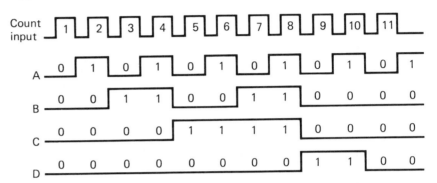

FIGURE 4–16

Timing Diagram for an Asynchronous Decade Up Counter

Synchronous Decade Down Counter

Synchronous decade counters are available, as are *decade down counters*. Decade down counters use four bits to count down from 9 through 0.

Like binary counters, decade counters divide the input frequency, but they divide it by ten, not two. Decade counters can be cascaded, as illustrated in Figure 4–17, to increase count capability. To read the number stored in the flip-flop, note that U_1 is the *least significant digit* (LSD) and U_4 is the *most significant digit* (MSD). Thus, we have

$$U_4 = 5 = 0101_2 \text{ (MSD)}$$

$$U_3 = 4 = 0100_2$$

$$U_2 = 8 = 1000_2$$

$$U_1 = 1 = 0001_2 \text{ (LSD)}$$

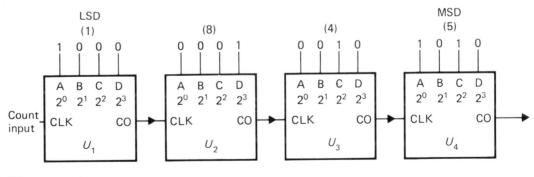

CO = carry out

FIGURE 4–17

Cascading Decade Counters to Increase Count Capability

or the decimal number 5481. Thus, the number of input pulses is decimal 5481, assuming that the counter was reset before starting.

COUNT CAPACITY

The maximum *count capacity* of a single decade counter is 9. Three stages of cascaded decade counters have a maximum count capacity of 999 [9 + (9 × 10) + (9 × 100)]. The maximum count of any number of stages can be found by this method. The maximum count of the four-stage decade counter in Figure 4–18, for example, is 9999 [9 + (9 × 10) + (9 × 100) + (9 × 1000)].

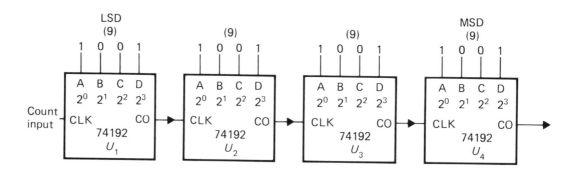

FIGURE 4–18

Maximum Count Capacity of Four-Stage Decade Counter

FREQUENCY DIVIDERS

The four-stage cascaded system shown in Figure 4–19 is used as a *frequency divider*. The output A of flip-flop U_1 is 1/10 of the input frequency and output D of U_4 is 1/10,000 of the input frequency. So, if a 1 MHz input is applied, the output D at U_4 will be 100 Hz. When counters are used as dividers, they are often called *scalers*. The waveforms for the 1 MHz input and the output A of the frequency divider circuit are also shown in Figure 4–19.

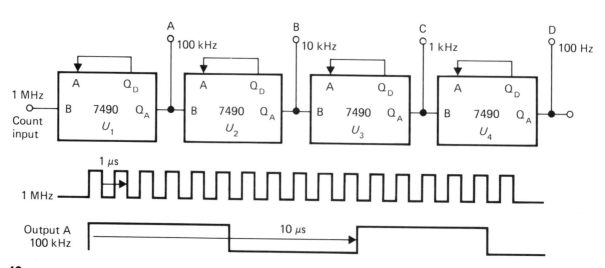

FIGURE 4–19

Frequency Divider Circuit and Timing Diagram

We saw in Chapter 3 that each JK flip-flop divides the input frequency by two. Decade counters divide by ten. Decade counters can be cascaded with binary counters to divide by a selected number. For example, if a decade counter is cascaded with a two flip-flop binary ripple counter, the division is by forty $[f/(10 \times 2 \times 2)]$. See Figure 4–20. That is, the output will be 1/40 of the input frequency. If the cascading is done with a four flip-flop binary counter, then the output will be 1/160 of the input frequency $[f/(10 \times 2 \times 2 \times 2 \times 2)]$. See Figure 4–21.

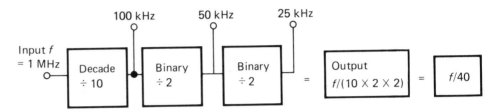

FIGURE 4–20

Divider Circuit with Output = 1/40 of Input Frequency

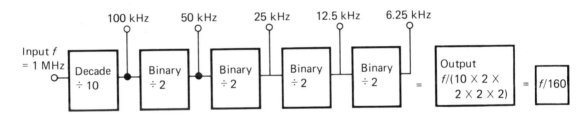

FIGURE 4–21

Divider Circuit with Output = 1/160 of Input Frequency

SELF-TEST EXERCISE 4–4

1. A (decade, binary) counter is a divide-by-two-and-divide-by-five counter combined in one IC to divide by ten.
2. A decade up counter counts from 0 through _____.
3. A decade down counter outputs 1000_2 and then outputs _____.
4. One decade counter divides its input by _____.
5. A decade counter is set to 0110_2. If ten more pulses are applied, its output is _____.
6. If a 3 MHz signal is applied, the output of a three-stage cascaded decade counter will be _____ Hz.
7. If five stages of a decade counter are cascaded, the maximum output count capability is _____.
8. A 5-bit binary counter has an output frequency that is _____ of the input frequency.

CONTROL PINS

Most complex ICs have *control pins*. The input control pin for an up/down counter is one example. The most common controls are the reset and set control pins.

The *reset control* is used to clear or zero the flip-flops (all the Q outputs are then zero). Often, we need to clear a counter before it starts the counting operation; so clear or reset is important, and all counters have this control.

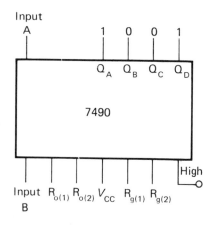

FIGURE 4–22

7490 Counter with Set-to-Nine Control Pins

Besides the reset control, some counters also have a *preset control*. This control allows the counter to be set to some value before it starts the counting operation. For example, the 7490 counter shown in Figure 4–22 has set-to-nine input control pins $R_{g(1)}$ and $R_{g(2)}$. When either of these input pins goes high, it sets the outputs Q_A, Q_B, Q_C, and Q_D equal to binary 1001, or decimal 9, as shown. The $R_{o(1)}$ and $R_{o(2)}$ clear control pins set all outputs to zero. For the 7490 counter to be used as a decade counter, the output Q_A is connected to input B. The pulses to be counted are then applied to input A.

The 74490 counter shown in Figure 4–23 has two separate 4-bit decade counters. Each counter in this IC has its own inputs, outputs, and control pins; but the two counters share the V_{CC} and ground (GND) connections. The two counters in the IC can be cascaded to increase count capability.

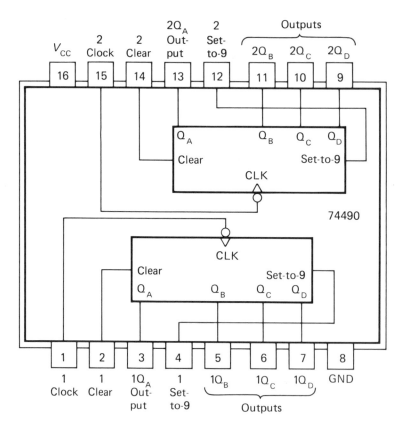

FIGURE 4–23

74490 Counter

Positive logic: High input to clear resets all four outputs low. High input to set-to-9 sets Q_A and Q_D high, Q_B and Q_C low.

Some counters, like the 74193 shown in Figure 4–24, have a *load control*. This control is used to load a binary number before the beginning of the count. The number is loaded through parallel inputs, as shown in the figure.

The 74193 counter also has *carry out* and *borrow out* pins. The carry out is connected to the count up input of the next flip-flop when the counter is cascaded to increase its capacity. The borrow out is connected to the count down input of the next flip-flop when the counter is cascaded for use as a down counter.

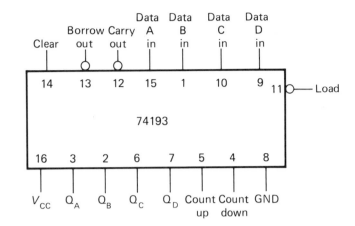

FIGURE 4–24

74193 Counter

MODULO *N* COUNTERS

Most counting applications can be handled by decade or binary counters. Some special applications require that a counter count only a few states repeatedly. The number of states that a counter has is called the *modulus* of the counter. A *modulo N counter* is a counter that has *N* states. Regular decade counters are modulo 10 counters. Depending on the application, some special modulo counters have been developed. Basically, these counters are binary or decade counters with modifications. For example, a modulo 13 counter counts from 0 to 12. It has thirteen states and counts to a maximum of decimal 12 before it recycles to zero.

  Do Experiment 4–2

SUMMARY

A binary counter, also called a ripple counter, is formed by cascading JK flip-flops. A binary counter outputs information in the pure binary format (8–4–2–1 for a 4-bit counter).

    The maximum decimal count for a binary counter is determined by the number of flip-flops in the counter. A 4-bit binary counter can count from 0 to 15. The number of pulses that can be counted is calculated by using 2^N, where *N* is the number of output bits. For example, a 4-bit counter can count 2^4, or 16, pulses. The maximum decimal number that can be represented by the binary output is $2^N - 1$. For example, an 8-bit counter can count $2^8 - 1$, or 255, items before it "turns over" to zero.

    A counter is an up counter when each input pulse increases the number stored in the counter by one. A counter is a down counter when each input pulse decreases the number stored in the counter by one. Counters are available that can count up and down, depending on the state of the control pins or where the count is applied.

    Flip-flops change state on the transition of the clock, either the leading or the trailing edge. If they change state on the zero-to-one transition, they are positive-edge triggered. If they change state on the one-to-zero transition, they are negative-edge triggered.

    In asynchronous counters, the individual JK inputs are not controlled by a single simultaneous clock pulse. In synchronous circuits, all elements are timed to a master timing signal or clock. Propagation delay can be minimized when synchronous counters are used.

A decade counter is one that outputs in binary form the numbers 0 through 9. It is also called a BCD counter. A decade counter recycles to zero after the ninth count. One decade counter is used for each decimal place in a number. Therefore, to represent the decimal number 5,493,306, seven decade counters would be needed. These same seven counters could count to 9,999,999 before they recycled to zero.

Counters can be used as frequency dividers. Binary counters divide by two; decade counters divide by ten. Combinations of flip-flops can be used to divide by any number desired. If a binary counter is used as a divider, the final output frequency (f_o) is $1/(2^N)$ multiplied by the frequency of the input (f_i). For a 100 Hz input frequency, the output of a 4-bit counter is 6.25 Hz: $f_o = 1/(2^N) \times f_i = 1/(2^N) \times 100 \text{ Hz} = 6.25 \text{ Hz}$.

Some special applications require a counter that recycles at a different point than the one at which a binary or a decade counter recycles. For these applications, a modulo N counter is used. A modulo 12 counter has twelve states; it recycles after the eleventh count.

CHAPTER 4
REVIEW EXERCISES

1. Which of the following functions is not a basic function of sequential logic circuits?

   a. make decisions
   b. generate timing pulses
   c. count
   d. produce automatic sequencing

2. The type of sequential logic circuit that operates from a clock is called a _____.

   a. counter
   b. one shot
   c. synchronous circuit
   d. combinational circuit
   e. asynchronous circuit

3. A binary up counter with flip-flops EDCBA (A = LSB) contains the number 00110. How many input pulses must be applied to obtain the contents 11000?

   a. 7
   b. 16
   c. 18
   d. 24

4. Another name for a decade counter is _____.

   a. binary counter
   b. BCD counter
   c. frequency divider
   d. shift register

5. What is the maximum decimal number that can be counted by a binary counter with twelve flip-flops?

a. 12
b. 24
c. 2047
d. 4095

6. How many decade counters are needed to count to the number 816?

a. 3
b. 4
c. 10
d. 12

7. What is the output frequency of the circuit in Figure 4–25?

a. 3.9 kHz
b. 6.25 kHz
c. 20 kHz
d. 50 kHz

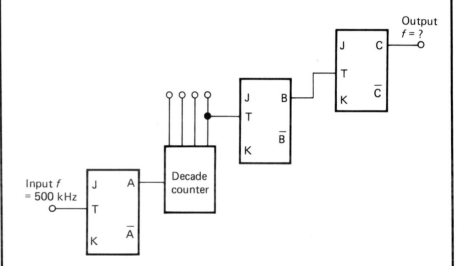

FIGURE 4–25

8. What circuit could be used to add and subtract input pulses?

a. shift registers
b. one shot
c. frequency divider
d. up/down counter

9. (Synchronous, Asynchronous) counter circuits change all output states A, B, C, and D at once.

10. The trailing edge of an input pulse is the transition from (1 to 0, 0 to 1).

11. A synchronous counter _____.

a. is slower than a ripple counter.
b. cannot count nonperiodic inputs.
c. changes state in a time equal to the propagation delay of one flip-flop.
d. is one in which one flip-flop toggles the next in sequence.

12. A binary up/down counter with flip-flops EDCBA (A = LSB) contains the number 10001. Five pulses are applied to the up count input. Twenty-four pulses are applied to the down count input. What is the new counter contents?

    a. 01110
    b. 11000
    c. 11101
    d. 11110

13. Synchronous and asynchronous counters can have any number of flip-flops and corresponding outputs. True or False?

14. Control pins override the clock; therefore, they are (synchronous, asynchronous) in operation.

15. Propagation delay adds up in (synchronous, asynchronous) counters.

16. How many states does a modulo 15 counter have?

17. What maximum decimal number could a modulo 15 counter represent?

EXPERIMENT 4–1 | Synchronous Binary Up/Down Counter

PURPOSE

This experiment is designed to demonstrate the operation of a basic binary counter. You will observe the counter as it operates as an up counter and as a down counter. You will use the load function to preset the counter to a specified value before the count begins.

EQUIPMENT

—1 logic probe ($Equip_1$)
—1 digital experimenter ($Equip_2$)
—1 dual-trace oscilloscope ($Equip_3$)
—1 SN74193 synchronous binary up/down counter (U_1)
—1 1000 Ω, 1/4 W resistor (R_1)

PROCEDURE

Step 1 Carefully wire the circuit shown in Figure 4E1–1. Be sure the power inputs are connected.

NOTE: The count or data input is from logic switch A. The reset or clear is from logic switch B. This IC has individual preset lines for each flip-flop. These lines are connected to the experimenter data switches. Outputs A, B, C, and D are connected to logic indicators L_1, L_2, L_3, and L_4.

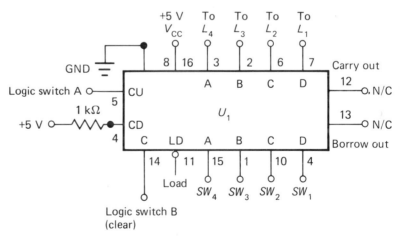

FIGURE 4E1–1

CU = count up
CD = count down
N/C = no connection

Step 2 Apply power.

Step 3 Reset the counter by depressing logic switch B.

Step 4 Apply pulses at the data input; use logic switch A while you observe the count sequence until the counter recycles again.

Check: What is the maximum decimal number that can be represented in binary with this counter?

15

Check: Is this a synchronous up or down counter?

up counter.

Step 5 Clear the counter while you hold the logic probe on pin 12.

Step 6 Depress logic switch A and count the number of pulses until the logic probe pulses low.

NOTE: The first time it will take only fifteen pulses because clear was the first count state.

Step 7 Depress logic switch A again until another low pulse is seen on the probe.

NOTE: It should take sixteen pulses. Every sixteenth input count pulse is output at pin 12 as an active low trigger pulse.

Step 8 Move the connections at logic switch A to a 1000 Hz clock on the experimenter. Connect channel B of the scope here.

Step 9 Connect channel A of the scope to carry out pin 12.

NOTE: The display on the scope should look like the one in Figure 4E1–2. This carry out pulse is connected to the count input of the next flip-flop when counters are cascaded to increase count capability.

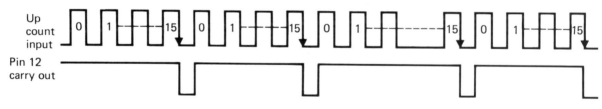

FIGURE 4E1–2

Check: Is the output at pin 12 active high or low?

low

Check: Is the clear input active high or low?

high

Step 10 Turn off the experimenter.

Step 11 Remove all wires connected to pins 4 and 5.

Step 12 Connect pin 4 to logic switch A.

Step 13 Connect pin 5 to V_{CC} through R_1.

Step 14 Turn on the experimenter.

Step 15 Clear the counter.

Step 16 Depress logic switch A while you observe the count operation with the logic indicators.

> **Check:**   Is this a synchronous up or down counter?
>
> down

ACTIVITY _____   Monitor pin 13 by using the logic probe.

NOTE: This borrow out pulse is connected to the count down input of the next flip-flop when counters are cascaded to increase count capability.

Step 17 Turn off the experimenter.

Step 18 Remove the logic switch connection at pin 14 and ground pin 14.

Step 19 Connect pin 11 to logic switch $\overline{B}$ and connect pin 5 to logic switch A.

Step 20 Ground pin 4 through R_1.

Step 21 Set data switches 1, 2, 3, and 4 to zero. Turn on the experimenter.

Step 22 Depress logic switch $\overline{B}$ and release.

ACTIVITY _____   Record the binary output word on the logic indicators: (DCBA)_____. (It should be 0000_2.)

Step 23 Set all data switches to one.

Step 24 Load this data word by depressing logic switch $\overline{B}$. (The logic indicators should show 1111_2.)

Step 25 Set the data switches to load the word 1010_2. Depress logic switch $\overline{B}$ and release.

NOTE: The logic indicators should now show the data word 1010_2. They should not have changed until the load input was triggered.

Step 26 Apply count pulses with logic switch A.

NOTE: The counter should count up from the preset value of 1010_2. It will recycle to 0000_2 and begin counting up again if sufficient additional pulses are applied.

ACTIVITY

This unique operation can be used to design a counter with any modulo needed. This circuit can be used as a modulo 5 counter—that is, a counter with five states. It simply needs one minor change to the circuit, which is to connect the carry out pin 12 to the load input instead of to the logic switch $\overline{B}$. Try it.

| EXPERIMENT 4–2 | # Operation of a Decade Counter |

PURPOSE

This experiment is designed to demonstrate the operation of a decade counter. You will build a divide-by-ten circuit from a decade counter.

EQUIPMENT

—1 digital experimenter (Equip$_1$)
—1 dual-trace oscilloscope (Equip$_2$)
—1 logic probe (Equip$_3$)
—1 SN7490 decade counter IC (U_1)

PROCEDURE

Step 1 Construct the circuit shown in Figure 4E2–1. Connect the power pins as shown.

NOTE: The counter input is from logic switch A. The reset is connected to data switch SW_1. The preset-to-nine is connected to data switch SW_2. The outputs are connected to the logic indicators.

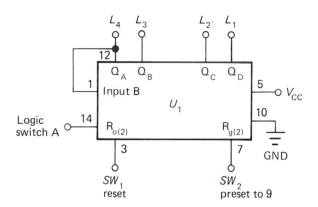

FIGURE 4E2–1

Step 2 Make sure that both SW_1 and SW_2 are in the zero position. Then, apply power.

NOTE: Some logic indicators may be lit.

Step 3 Set SW_1 to one and back to zero.

ACTIVITY

Check the logic indicators. The counter should have reset to zero. If it has not, use the logic probe to check the power pins for V_{CC} and ground. Then, make sure a high pulse is applied to pin 3 when SW_1 is moved high to low.

Step 4 Apply pulses by depressing and releasing logic switch A. Observe the count sequence after every pulse.

ACTIVITY

Describe what happens after the ninth count (1001_2).

Step 5 Clear the counter again by moving data switch SW_1 to one for a moment and returning it to zero.

Step 6 Move data switch SW_2 to one for a moment and return it to zero while you observe the count.

NOTE: The counter should have set to nine; the output should be 1001_2 (9). This operation does not generally have much application.

Step 7 Remove logic switch A from pin 14.

Step 8 Apply a 1 Hz clock to input pin 14.

ACTIVITY

Watch the counter count for a while to become familiar with the counting sequence.

NOTE: Outputs A, B, C, and D can be observed on the oscilloscope if you increase the clock frequency to 1000 Hz or more.

Step 9 Turn off the experimenter.

Step 10 Remove the jumper from pin 12 to pin 1. Add a jumper from pin 11 to pin 14.

Step 11 Remove the clock connection at pin 14.

Step 12 Connect logic switch A to pin 1.

NOTE: You now have a divide-by-ten circuit. The output is at Q_A, logic indicator L_4.

Step 13 Turn on the experimenter.

Step 14 Clear the counter.

Step 15 Depress logic switch A five times; L_4 should light.

Step 16 Depress logic switch A five more times; L_4 should go out.

NOTE: The timing diagram for this operation is shown in Figure 4E2–2. The frequency of output Q_A, logic indicator L_4 is 1/10 of the input frequency.

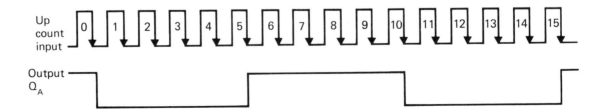

FIGURE 4E2–2

Step 17 Turn off the experimenter.

Step 18 Remove logic switch A from pin 1.

Step 19 Apply a 1000 Hz clock to input B pin 1 and also connect channel A of the scope to this point.

Step 20 Connect channel B of the scope to the 1000 Hz clock.

Step 21 Turn on the experimenter.

> *NOTE:* The scope should display waveforms like the ones shown in the timing diagram in Figure 4E2–2.

Shift Registers

OBJECTIVES

After studying this chapter, you will be able to:

1. Describe the construction and operation of a simple shift register.
2. Describe several applications of shift registers.
3. Demonstrate the number of steps required to input a given serial word into a shift register.
4. Demonstrate the number of steps required to input a given parallel word into a shift register.
5. Multiply or divide by powers of two a given word through the use of a shift register.
6. Draw the truth table for an N-bit ring counter (N is the number of bits).
7. Describe the advantages of MOS-type registers.
8. List the four classifications of shift registers.
9. Define the terms *single-rail input* and *double-rail input*.

INTRODUCTION

Digital information sometimes must "wait" until it is needed somewhere—that is, it must be held in a particular location temporarily and then be shifted to where it is required. A shift register is a special circuit that not only holds and shifts information but also divides clock signals for special timing applications. This chapter describes shift registers and examines their basic functions.

TYPES OF SHIFT REGISTERS

A *shift register* is a memory or sequential logic system that consists of flip-flops or MOS dynamic cells. A shift register transfers data from one register to another as many times as necessary. Shift registers are often used when temporary storage of data is needed. For example, if addition is to be performed on a column of numbers, as each number is added to the next, the partial sum can be momentarily stored in a register and then added to the next number in the column. Also, in some cases, the data in a storage register waits until another part of the circuit requires it.

We can see, then, that shift register operation is like that of a bus stop: Data accumulates in a register and waits to be picked up by another part of the circuit. The storage elements are connected so that stored bits can be shifted left or right from one element to another. Shift register circuits are widely used as counters.

Shift registers are available as MSI circuits in an IC chip with various capabilities, including serial-in/serial-out, serial-in/parallel-out, or a combination of these capabilities. A *universal shift register*, for example, is capable of right or left shift and both serial and parallel data entry and output. This type of register is generally used when only a few stages are needed because it requires many input and output pins.

TTL shift registers are available in an IC chip with up to ten stages. The small size of the MOSFET structure makes MOS devices with 2000 to 4000 stages common and allows many words to be stored in a small space. Some ICs—microprocessor ICs, for example—contain internal shift registers that are only a small portion of their total circuitry.

Shift registers are also used between two systems where timing and/or mode variations exist. They are known as *buffer registers*.

Serial Shift Register

In a *serial shift register*, data is entered serially, one bit at a time, until the desired number of bits has been entered. In this simplest type of shift register, all the storage elements are triggered at the same time by a clock pulse. The result is that data present in a flip-flop *prior* to a clock pulse is shifted to the next flip-flop *with* the clock pulse. The number of clock pulses required to enter a binary word serially is equal to the number of bits in the word. So, for example, a 4-bit word requires four clock pulses.

Serial shift registers have D or JK inputs to the first stage. Usually, JK flip-flops are used because they are more versatile than D-type flip-flops. A typical serial shift register is shown in Figure 5–1. Four JK flip-flops are cascaded by connecting the outputs of one flip-flop to the JK inputs of the next. The serial data and its complement are fed to the JK input of the first flip-flop. All clock inputs are tied together to a common clock input. In this way, each clock pulse shifts the stored data to the right by one location.

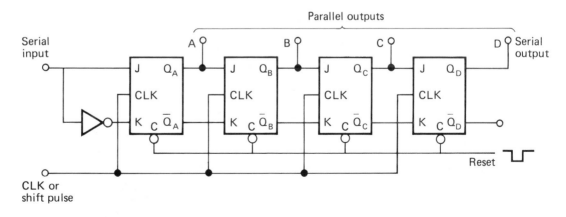

FIGURE 5–1

Serial Shift Register with JK Inputs

The waveforms in the timing diagram shown in Figure 5–2 indicate how the input data is moved into the serial shift register one bit at a time. Notice that data pulse D_1, which enters first, moves to the right by one location for each clock pulse. The 4-bit data word 0101_2 is available as a parallel output after four clock pulses. All data shifts out of the register after eight clock pulses. Usually, the clock or shift pulses are common to all the circuits in the system so that any serial data available will be synchronized to the clock.

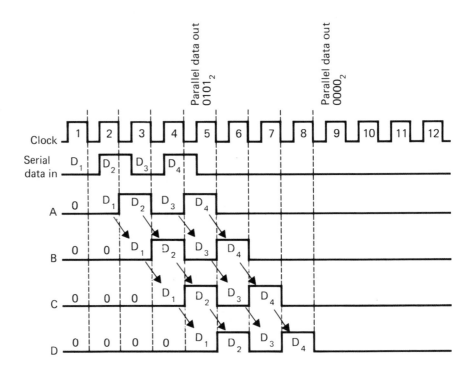

FIGURE 5–2

Serial Shift Register Timing Diagram

As many flip-flops as are needed are cascaded together in shift registers. Many digital systems have long shift registers to store a number of bits. Although shift registers can be built with JK or D flip-flops, MSI shift registers are the most practical. These registers are much more capable than the register shown in Figure 5–1 because they have shift left/right, parallel in/out, and serial in/out controls, or some combination of these capabilities, together in one IC.

SELF-TEST EXERCISE 5–1

1. How many pulses are needed to store an 8-bit word serially into an 8-bit shift register?
2. Shift registers can be used for parallel-to-serial and serial-to-parallel data conversions. True or False?
3. A _____ register transfers data from one register to another.
4. Shift registers can be used as _____ between two systems where timing and/or mode variations exist.
5. _____ -type shift registers are the most versatile.

6. MSI shift registers are more versatile and practical than flip-flop registers. True or False?

STOP   Do Experiment 5–1

Parallel Shift Register

In a *parallel shift register* a complete data word is entered at once. In the 74195 parallel shift register shown in Figure 5–3A, data available at data inputs A, B, C, and D are loaded on the leading edge of the clock pulse (positive transition) when the shift/load control is low. During parallel loading, the serial data input is inhibited.

      The truth table for this register is shown in Figure 5–3B. Notice, from the truth table, that this register is also capable of serial load and shift.

SINGLE- AND DOUBLE-RAIL INPUTS

In a *single-rail input register*, such as the one in Figure 5–1, the JK inputs are tied together with an inverter that supplies the K input. Only Q is output; no $\overline{Q}$ is available.

      In a *double-rail input register*, the JK inputs are available as separate data inputs. See, for example, pins 2 and 3 of the 74195 register in Figure 5–3.

CHARACTERISTICS OF SHIFT REGISTERS

Table 5–1 lists some common TTL and CMOS shift registers and their specific characteristics. Details of these characteristics are discussed next.

TABLE 5–1

Characteristics of Shift Registers

Length (Stages)	Type	TTL or CMOS	Parallel Outputs	Direction	Parallel Load	Clear
8	7491	TTL	No	Right	No	No
4	7494	TTL	No	Right	Preset only	Yes
4	7495	TTL	Yes	Right/left	Synchronous	No
4	74C95	CMOS	Yes	Right/left	Synchronous	No
5	7596	TTL	Yes	Right	Preset only	Yes
8	74164	TTL	Yes	Right	No	Yes
8	74C164	CMOS	Yes	Right	No	Yes
8	74165	TTL	No	Right	Asynchronous	Yes
8	74C165	CMOS	No	Right	Asynchronous	Yes
8	74166	TTL	No	Right	Synchronous	Yes
4	74194	TTL	Yes	Right/left	Synchronous	Yes
4	74195	TTL	Yes	Right	Synchronous	Yes
4	74C195	CMOS	Yes	Right	Synchronous	Yes

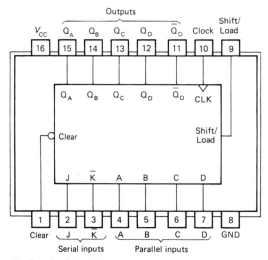

A. Pin Assignments

Inputs									Outputs				
Clear	Shift/ Load	Clock	Serial		Parallel				Q_A	Q_B	Q_C	Q_D	$\overline{Q}_D$
			J	$\overline{K}$	A	B	C	D					
L	X	X	X	X	X	X	X	X	L	L	L	L	H
H	L	↑	X	X	a	b	c	d	a	b	c	d	$\overline{d}$
H	H	L	X	X	X	X	X	X	Q_{A_o}	Q_{B_o}	Q_{C_o}	Q_{D_o}	$\overline{Q}_{D_o}$
H	H	↑	L	H	X	X	X	X	Q_{A_o}	Q_{A_o}	Q_{B_n}	Q_{C_n}	$\overline{Q}_{C_n}$
H	H	↑	L	L	X	X	X	X	L	Q_{A_n}	Q_{B_n}	Q_{C_n}	$\overline{Q}_{C_n}$
H	H	↑	H	H	X	X	X	X	H	Q_{A_n}	Q_{B_n}	Q_{C_n}	$\overline{Q}_{C_n}$
H	H	↑	H	L	X	X	X	X	$\overline{Q}_{A_n}$	Q_{A_n}	Q_{B_n}	Q_{C_n}	$\overline{Q}_{C_n}$

H = high, L = low
X = irrelevant
↑ = low-to-high transition
a, b, c, d, = level of steady state input at A, B, C, D, respectively, before indicated steady state input conditions were established
Q_{A_n}, Q_{B_n}, Q_{C_n} = level of Q_A, Q_B, Q_C, respectively, before most recent transition of clock

B. Truth Table

FIGURE 5–3

74195 Parallel Shift Register

Clear. As with all clear control pins, the clear control for a shift register resets all flip-flops to zero. Most shift registers have this control, but when it is not available, a register is cleared by entering all zeros at the data input(s) until all Q lines are low.

Preset. Parallel load shift registers with a preset-only capability must be cleared to all zeros and the register reloaded through the preset inputs. That is, a one in an individual stage cannot be changed to a zero. True parallel registers allow updating at any time without.clearing first.

Synchronous Load. In synchronous (clocked) load devices, loading always occurs with the leading or trailing edge of the clock pulse. Most shift registers clock on the leading edge of the clock pulse (positive transition).

Asynchronous Load. In asynchronous load devices, loading of data at the parallel inputs occurs on the high-to-low transition of the shift/load input. Since the operation is independent of the levels of the clock, clock inhibit, or serial inputs, it is considered asynchronous. For an example of a parallel asynchronous load register, see the 74165 shown in Figure 5–4.

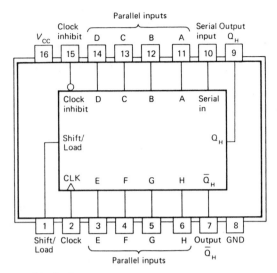

A. Pin Assignments

Inputs					Internal outputs		Output
Shift/ Load	Clock inhibit	Clock	Serial	Parallel A ... H	Q_A	Q_B	Q_H
L	X	X	X	a ... h	a	b	h
H	L	L	X	X	Q_{A_o}	Q_{B_o}	Q_{H_o}
H	L	↑	H	X	H	Q_{A_n}	Q_{G_n}
H	L	↑	L	X	L	Q_{A_n}	Q_{G_n}
H	H	X	X	X	Q_{A_o}	Q_{B_o}	Q_{H_o}

B. Truth Table

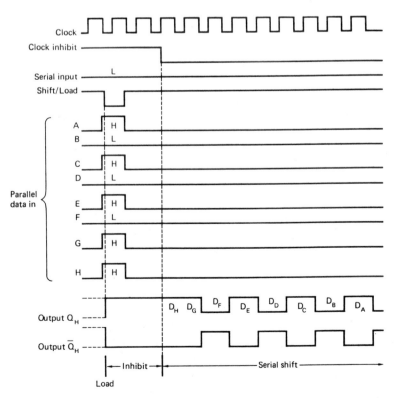

FIGURE 5–4

74165 Parallel Asynchronous Load Register

C. Timing Diagram

Parallel Outputs. If parallel outputs are not available, a serial-only register is used. For an example of a serial-only register, see the 7491 shown in Figure 5–5. Parallel-in/parallel-out registers are very useful in applications where serial-to-parallel and parallel-to-serial data conversions are required.

Classifications of Shift Registers. Shift registers are characterized as serial-in/serial-out (SISO), serial-in/parallel-out (SIPO), parallel-in/serial-out (PISO), and parallel-in/parallel-out (PIPO).

Parallel Enable. Also called the *mode control*, the parallel enable pin controls how data is to be entered (serial or parallel). It should not be allowed to change states during clocking; a 10 nanosecond delay after the clock pulse is needed before the mode can be switched. This input is

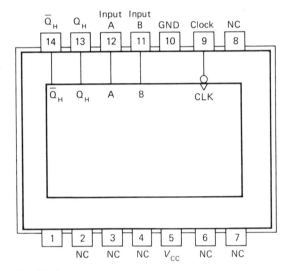

A. Pin Assignments

Inputs at t_n		Outputs at t_{n+8}	
A	B	Q_H	$\overline{Q}_H$
H	H	H	L
L	X	L	H
X	L	L	H

H = high, L = low
X = irrelevant
t_n = reference bit time, clock low
t_{n+8} = bit time after 8 low-to-high transitions

B. Truth Table

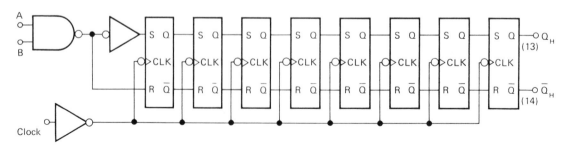

C. Functional Block Diagram

FIGURE 5–5

7491 Serial-Only Register

especially useful when a single register is to be used for both serial and parallel processing.

Clock. Two clock inputs are usually provided if parallel and serial entries are made. They are clocked at different rates to allow multiple use of single registers.

Direction. Shifting data to the left or the right by a specific number of positions is equivalent to multiplying or dividing the stored number by a factor.

SELF-TEST EXERCISE 5-2

1. With single-rail inputs to a shift register, only a binary one can be shifted. True or False?
2. In double-rail shift registers, both the _____ inputs are available as separate data inputs.
3. All shift registers are asynchronous. True or False?
4. The shift register classification PIPO means _____.
5. The 74165 register has parallel inputs for (synchronous, asynchronous) operation.
6. Two clock inputs may be available for some shift registers. True or False?
7. Shift registers can be used to multiply and _____ stored data.

STOP   Do Experiment 5-2

APPLICATIONS OF SHIFT REGISTERS

Scalers

Shift registers have many applications besides data storage. We already know that they can be used as parallel-to-serial and serial-to-parallel data converters. They can also be used as *scalers* to multiply and divide the data stored. Shifting left is equivalent to multiplying the data by some power of two. Shifting right is equivalent to dividing the data by some power of two. Examples of register multiplication and division are shown in Figure 5-6.

16	8	4	2	1	
0	0	1	0	0	Initial = 2
0	0	1	0	0	First shift left = 4
0	1	0	0	0	Second shift left = 8
1	0	0	0	0	Third shift left = 16

A. Multiplication by Shifting Left

16	8	4	2	1	
1	1	0	0	0	Initial = 24
0	1	1	0	0	First shift right = 12
0	0	1	1	0	Second shift right = 6
0	0	0	1	1	Third shift right = 3

B. Division by Shifting Right

FIGURE 5-6

Shift Register Multiplication and Division

Ring Counters

Shift registers are also used as *ring counters* or *sequencers*. In a ring counter, the data is shifted "around" the register. That is, when it reaches the last Q out, it moves back to Q_A. See Figure 5–7. This shifting of data results in equally spaced timing pulses that are required for control of different operations in many digital circuits. Shift registers connected as a ring counter are shown in Figure 5–8.

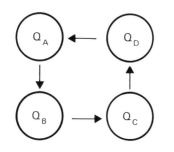

FIGURE 5–7

Ring Counter

FIGURE 5–8

Shift Registers Connected as a Ring Counter

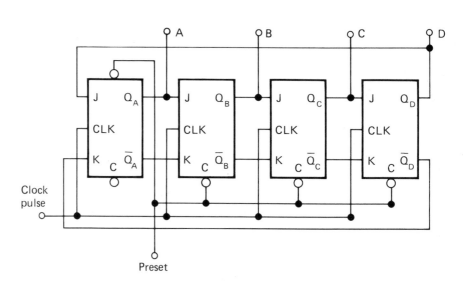

A 4-bit ring counter has four states. The truth table and timing diagram for a simple 4-bit ring counter are shown in Figure 5–9. A 5-bit ring counter has an additional flip-flop to pass through before recycling; therefore, it has five states. Thus, we may conclude that a ring counter has the same number of states as it has flip-flops or bits (since each flip-flop outputs one bit).

FIGURE 5–9

4-Bit Ring Counter

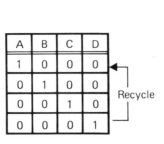

A	B	C	D
1	0	0	0
0	1	0	0
0	0	1	0
0	0	0	1

Recycle

A. Truth Table

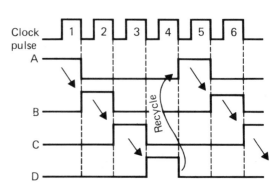

B. Timing Diagram

SUMMARY

A shift register is a temporary data storage area made up of flip-flops or MOS dynamic cells. MOS registers are very small, allowing one IC chip to contain thousands of individual stages and hold thousands of bits of data at a time.

Shift registers are available with serial or parallel data entry and output or combinations of both. They are classified as SISO, PIPO, SIPO, and PISO. One clock pulse is required for each digit in a serial entry register. Only one clock pulse is needed to enter a complete data word in a parallel entry register.

Shift registers can be used for arithmetic functions as well as for data storage. A binary word stored in a shift register can be shifted left to multiply by two or shifted right to divide by two. For example, the binary number 01101010 (106) shifted right one position is 00110101 (53); shifted left one position, it is 11010100 (212).

CHAPTER 5
REVIEW EXERCISES

1. Shift registers can be constructed with D-type flip-flops. True or False?

2. Shift registers are used for: (Check all that apply)

   a. temporary storage
   b. to multiply and divide by powers of two
   c. parallel-to-serial data conversion
   d. none of the above

3. How many clock pulses are needed to load a 10-bit word *serially* in a 10-bit shift register?

   a. 100
   b. 5
   c. 10
   d. 15
   e. 1

4. How many clock pulses are needed to load a 16-bit word in a parallel register?

   a. 16
   b. 15
   c. 1
   d. 10
   e. 8

5. If the data in Figure 5–10 is shifted left three times, the arithmetic function performed is:

   a. addition
   b. subtraction
   c. multiplication
   d. division
   e. none of the above

FIGURE 5–10

MSB						LSB
0	0	0	1	0	0	1

6. How many bits can be stored in a shift register with five flip-flops?

   a. 32
   b. 31
   c. 5
   d. 4
   e. 1

7. A ring counter is an application of a:

   a. binary counter
   b. shift register
   c. BCD counter
   d. modulo 5 counter
   e. none of the above

8. How many times must the data in Figure 5–11 be shifted to multiply the data by four?

   a. two times right
   b. four times left
   c. two times left
   d. four times right

FIGURE 5–11

MSB   LSB

0	0	0	0	1	1	1	1

9. The main advantage(s) of a MOS shift register is (are):

   a. small size
   b. high density
   c. high speed
   d. low power consumption
   e. none of the above

10. What classification of shift register is SISO?

11. Explain the term *double rail*.

12. How many states are possible when a 7-bit ring counter is used?

13. Draw the truth table for the ring counter mentioned in Question 12.

EXPERIMENT 5–1

Operation of a Serial Shift Register

PURPOSE

This experiment is designed to show the operation of a simple shift register that is built from individual JK flip-flops. You will learn to appreciate registers in IC form. They are very easy to wire up because all interconnected wiring is internal to the chip.

EQUIPMENT

—1 digital experimenter (Equip$_1$)
—2 SN7476 IC JK flip-flops (U_1, U_2)
—1 SN7404 IC inverter gate (U_3)

PROCEDURE

Step 1 Connect $+5$ V and ground properly to each IC.

Step 2 Construct the 4-bit serial shift register as shown in Figure 5E1–1.

NOTE: Logic switch $\overline{A}$ supplies the clock input, and logic switch $\overline{B}$ is connected to the clear control of the shift register. The serial data input is from data switch SW_1 and is fed to J and through the inverter to K of the first flip-flop. The four outputs are connected to the four LED indicators.

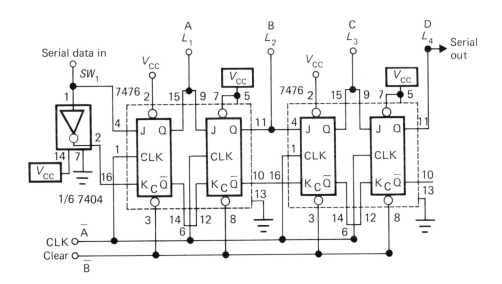

FIGURE 5E1–1

ACTIVITY

Check the wiring before you apply power.

Step 3 Apply power to the circuit.

Step 4 Clear the shift register by using logic switch $\overline{B}$.

Step 5 Set data switch SW_1 to binary one.

Step 6 Apply clock pulses to the circuit with logic switch $\overline{A}$.

> **Check:** What is the output of the shift register after every pulse for four shift pulses?
>
> pulse 1: ABCD = 1000_2
> pulse 2: ABCD = 1100_2
> pulse 3: ABCD = 1110_2
> pulse 4: ABCD = 1111_2

NOTE: The timing diagram for this operation, as shown in Figure 5E1–2, has seven clock pulses. The serial data in goes low before the clock pulse 5.

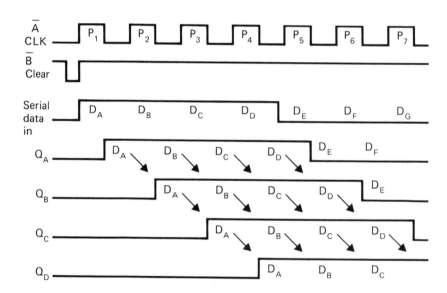

FIGURE 5E1–2

Step 7 Clear the shift register again and set data switch SW_1 alternately to one and zero.

> **Check:** What is the output of the shift register after every pulse for four shift pulses?
>
> pulse 1: ABCD = 1000_2
> pulse 2: ABCD = 0100_2
> pulse 3: ABCD = 1010_2
> pulse 4: ABCD = 0101_2

NOTE: From these steps, you can see that the data is shifted right by one position for each clock pulse. The four bits of serial data are available in parallel form through the outputs of each flip-flop after four pulses. The least significant bit is shifted out first.

Step 8 Continue to clock the register with data switch SW_1 at zero.

NOTE: All the data should be shifted out of the register after a total of eight pulses have been applied:

$$\text{pulse 5: ABCD} = 0010_2$$

$$\text{pulse 6: ABCD} = 0001_2$$

$$\text{pulse 7: ABCD} = 0000_2$$

$$\text{pulse 8: ABCD} = 0000_2$$

See the timing diagram for this operation, which is shown in Figure 5E1–3.

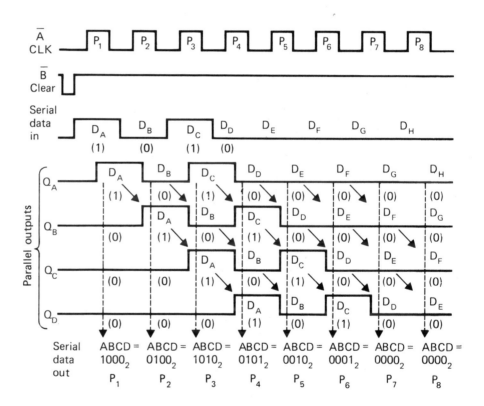

FIGURE 5E1–3

EXPERIMENT 5–2 | Operation of a Ring Counter

PURPOSE

This experiment is designed to show the operation of a simple ring counter. You will build one from a shift register IC chip. You will also draw a timing diagram of the operation of the ring counter.

EQUIPMENT

—1 digital experimenter (Equip$_1$)
—1 SN74194 IC 4-bit universal shift register (U_1)
—1 SN7422 IC NAND gate (U_2)
—1 SN7404 IC inverter (U_3)

PROCEDURE

Step 1 Connect the circuit as shown in Figure 5E2–1.

NOTE: Switch SW_1 should be high and SW_2 should be low to allow a shift right mode of operation. Switch SW_3 is the serial shift right data input; it should be low at this time.

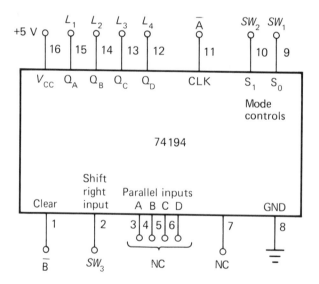

FIGURE 5E2–1

Step 2 Apply power. If any output logic indicators are on, clear the register by depressing $\overline{B}$.

Step 3 Serially, input the first bit by setting switch SW_3 high. Then, depress logic switch $\overline{A}$ once.

Step 4 Disconnect pin 2 from switch SW_3. Connect pin 2 to pin 12.

Step 5 Depress $\overline{A}$ several times to observe the ring counter operation.

> *NOTE:* Pin 12 is used to feed the last bit back to the serial shift right input as it is shifted out.

Step 6 Disconnect pin 11 from $\overline{A}$. Connect pin 11 to the experimenter clock.

> *NOTE:* The clock must be set to 1 Hz so that you can see the data shift.

Step 7 Connect the circuit as shown in Figure 5E2–2.

> *NOTE:* This self-starting, self-correcting circuit eliminates the need to set the first bit. Be sure power is connected to all ICs, SW_1 is high, SW_2 is low, and the 1 Hz clock is connected to pin 11.

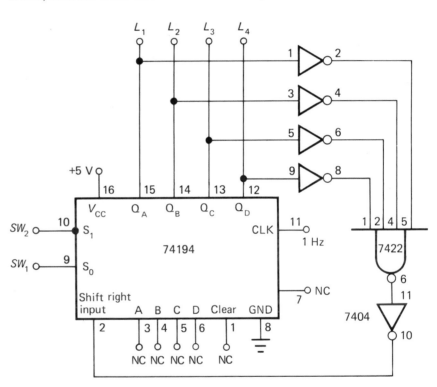

FIGURE 5E2–2

Step 8 Apply power.

> *NOTE:* This circuit will correct itself until the ring counter operation is correct and continual regardless of the beginning output.

ACTIVITY

Draw a timing diagram for this operation. Include ten or more state changes.

Step 9 Momentarily switch SW_2 to high and then to low to induce an error.

> *NOTE:* The error will be corrected automatically. Figure 5E2–3 shows the timing diagram for this circuit when an error was induced after clock pulse 3.

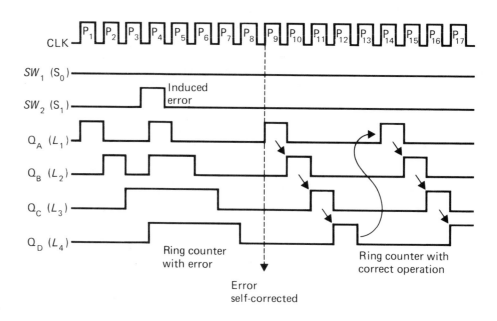

FIGURE 5E2–3

Step 10 Connect the circuit as shown in Figure 5E2–4.

> *NOTE:* The shift left input is used in this circuit. Mode switch SW_1 must be low and mode switch SW_2 must be high for shift left. The clock input, pin 11, can be connected to A for manual clocking or to a 1 Hz clock for automatic sequencing.

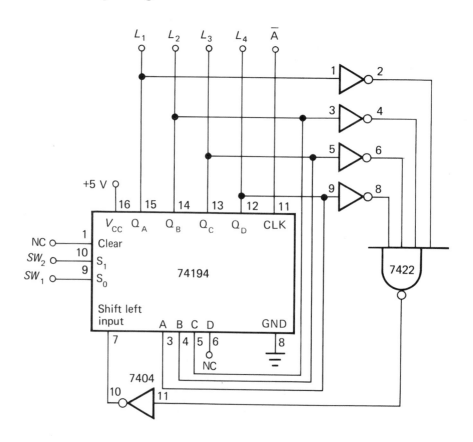

FIGURE 5E2–4

NOTE: The parallel outputs Q_B, Q_C, and Q_D are connected to the preceding parallel inputs: Q_B to A, Q_C to B, and Q_D to C. The shift left input is connected to the output of the inverter. Figure 5E2–5 shows the timing diagram for this self-starting, self-correcting circuit.

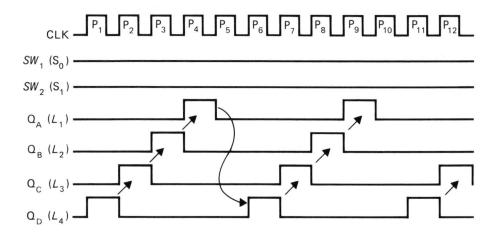

FIGURE 5E2–5

Check: How would this circuit be modified if an AND gate were used instead of the SN7422 NAND gate?

Inverter gate pins 10 and 11 would not be needed.

Clock Circuits

OBJECTIVES

After studying this chapter, you will be able to:

1. Describe the function of a master clock.
2. Build a simple astable multivibrator, monostable multivibrator, or crystal oscillator, given the parameters and a schematic.
3. Draw the waveform for a given clock frequency, specifying time period and labeling transitions as leading or trailing edge.
4. Select the proper type of oscillator for precision timing.
5. Define the origin of a subordinate clock.
6. Define an astable multivibrator and a monostable multivibrator and explain how each circuit operates.
7. Name the circuit parameter that determines the maximum operating clock frequency selected for a digital circuit.
8. Draw the waveform for a two-phase clock.
9. Determine the effect that increasing or decreasing capacitance has on the output pulse width of a monostable or an astable multivibrator circuit.
10. Describe the purpose of an IC buffer gate when it is used in an oscillator circuit.
11. Name the most popular timing circuit device (as cited in this chapter).
12. Define setup time and hold time.

INTRODUCTION

As the "supervisors" of digital circuits, clocks give the commands and determine the speed with which work will be performed by other circuits in the system. Clocks ensure that everything is completed at the proper time in the sequence of events. This chapter discusses various types of clock circuits and their operation.

MASTER CLOCK

The heart of computer and digital systems is the timing circuitry called the *master clock*. Most digital systems operate as synchronous sequential systems, where the sequence of operations that takes place is synchronized or controlled by a master clock signal. ICs are available with all of the parts of a system except the crystal or resistor–capacitor (RC) network and a few small components. A master clock generates periodic pulses that are distributed to all parts of the system. Each pulse is a reference waveform that determines the speed at which the system operates and how long it takes to perform an operation.

The master clock signal synchronizes all of the logic in a system. In particular, operations in the system are made to take place at times when the clock signal is making a transition from zero to one (rising or positive-going edge) or from one to zero (falling or negative-going edge). See Figure 6–1.

FIGURE 6–1

Master Clock Pulse

The master clock is an oscillator. It operates on a frequency determined by resistor–capacitor networks or, for precision timing, by crystal-controlled oscillators. It may be on a microprocessor chip or may be external. (Most microprocessors have an internal clock that needs only crystal or RC components to operate.) It emits pulses at a fairly high frequency—in the megahertz (MHz) range. These pulses refresh dynamic RAMs as well as synchronize the logic signals. (RAMs are discussed in Chapter 10.)

The timing of logic signals is accomplished through the use of gates that cannot switch until the clock pulse is received on the clock input. The data usually reach the inputs at a moment between clock pulses, but cannot be transferred until the clock pulse transition, as occurs in shift registers (see Chapter 5).

The frequency selected for the clock pulses is determined by the time required for the flip-flops and gates in a circuit to respond to level changes initiated by the clock pulses—that is, by the propagation delay of the various logic circuits. The master clock must be efficient long enough to perform the required operations in a given cycle, but not so long as to waste time. Note that this time is in the nanosecond (ns) range.

SUBORDINATE CLOCKS

Clock pulses are often needed that are shifted in phase from the master clock or that have pulse widths or repetition rates that differ from the basic pulse train. These clocks are called *subordinate clocks* because they are derived from the master clock and are not independent pulse generators.

Multiphase clocks are needed for memory circuits. These clocks are also derived from the master clock. These subordinate clocks may be on a memory chip, or, when subordinate clock functions for separate circuitry are required, special ICs may be used.

SELF-TEST EXERCISE 6-1

1. Digital systems are made to operate as synchronous sequential systems by a _____.
2. Operations in a system take place during the _____ of a clock pulse.
3. The master clock is an _____ with a frequency determined by either a crystal or an *RC* network.
4. Clock frequencies in the gigahertz (GHz) range are common. True or False?
5. _____ of the various flip-flops and gates determines the frequency selected for the master clock.
6. Clocks derived from the master clock are called _____ clocks.
7. Clock pulses are perfect square waves. True or False?

IC CLOCK CIRCUITS

Astable Multivibrator

An *astable multivibrator* may be defined as a free-running circuit that has two momentarily stable states, between which it continuously alternates, remaining in each state for a period controlled by circuit parameters and switching rapidly from one state to the other. Let us examine what this definition means.

First, it means that an astable multivibrator is high or low for a period of time that is controlled by the resistor–capacitor (*RC*) time constant. That is, it stays high or low for that period of time and then changes state. Second, it means that, while the astable multivibrator is in one of these states, it is fairly stable. Third, it means that it continues to function (is free-running) as long as power is applied to the circuit.

A discrete astable circuit is seldom used in modern equipment. The astable multivibrator shown in Figure 6–2 is constructed from integrated circuit inverter gates. The frequency of oscillation for this circuit depends on the values of resistance (*R*) and capacitance (*C*). The resistors provide the charge paths for the capacitors and bias the inverters. The frequency (*f*) for this circuit is approximately equal to $1/2.2\,RC$. The circuit produces a square wave output. A buffer should always be used to isolate the load from the frequency-determining components. This type of astable multivibrator is fairly accurate, but its accuracy is degraded with circuit age as well as other factors. Therefore, for precision timing, a more accurate and faster clock may be needed.

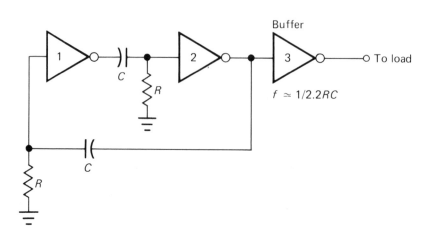

FIGURE 6–2

Astable Multivibrator with Two *RC* Networks

Figure 6–3 shows a crystal-controlled astable multivibrator. This circuit is modified to oscillate at the frequency of the crystal (XTAL). The crystal is a thin slab of quartz, cut and ground to a thickness at which it will vibrate at the desired frequency when it is supplied with energy. It is encased in a metal can that has two output pins where the necessary RC components are connected. The RC components are selected to oscillate near the crystal frequency. The frequency of operation for a particular crystal is set and cannot be changed. This type of circuit is much more accurate than the conventional astable circuit. Therefore, it is often used in applications where a stable and accurate clock is needed—in computers, for example, where precision clocking is essential. High-speed gates should also be used.

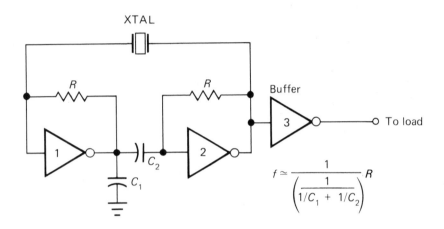

FIGURE 6–3

Crystal-Controlled Astable Multivibrator

The astable multivibrator shown in Figure 6–4 has only one RC network. The resistor value is critical and must be below 300 Ω for proper operation. The frequency of oscillation for this circuit is approximately equal to 1/3RC.

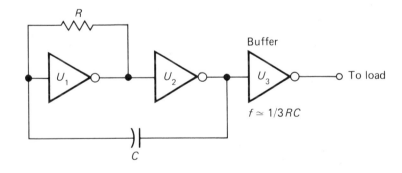

FIGURE 6–4

Astable Multivibrator with One RC Network

SELF-TEST EXERCISE 6–2

1. Define the term *astable multivibrator*.
2. Increasing the capacitance in an astable multivibrator _____ the frequency of oscillation.
3. Increasing the capacitance in an astable multivibrator _____ the time period of the clock pulse.
4. The frequency of oscillation of an astable multivibrator is determined by the _____ components or _____ selected.

5. A _____ should always be used to isolate the load from the frequency-determining components in an astable multivibrator.
6. The astable multivibrator shown in Figure 6–5 has an output frequency of _____ Hz.

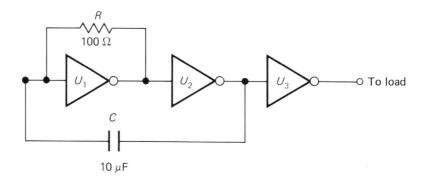

FIGURE 6–5

7. *RC* components for a _____ -controlled oscillator are selected to oscillate near the _____ frequency.

STOP Do Experiment 6–1

Monostable (One-Shot) Multivibrator

A *monostable*, or *one-shot*, *multivibrator* is a circuit that produces a fixed-duration output pulse each time it receives an input trigger. The duration of the pulse is usually controlled by external components. This type of circuit is used for delay and other forms of irregular timings.

As shown in the timing diagram in Figure 6–6, an input trigger pulse applied to a one-shot multivibrator results in an output pulse whose leading edge occurs at the same time as the input pulse but whose trailing edge occurs anytime before or after the end of the input pulse. The cascading of one-shot circuits allows a variety of sequential logic operations.

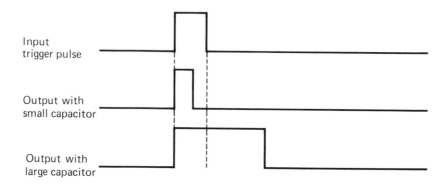

FIGURE 6–6

One-Shot Multivibrator Timing Diagram

IC One-Shot Multivibrator. Most one-shot circuits in use today are in integrated circuit form. Although its operation is virtually identical to a discrete component multivibrator circuit, the *IC one-shot multivibrator* is easier to use and much less complicated to troubleshoot.

Figure 6–7 shows the logic symbol used to represent IC one-shot multivibrators. Although the logic symbol appears to be an IC with two additional gates attached, it is actually one IC with the inputs shown. The only external components needed are the resistor and capacitor.

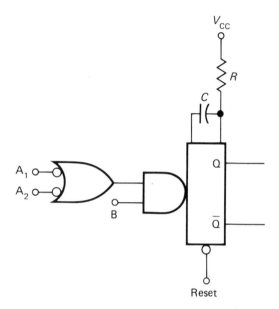

FIGURE 6–7

IC One-Shot Multivibrator Logic Symbol

Some IC monostable multivibrators, like the 74121 shown in Figure 6–8, also have an internal resistor (R_{int}). An output pulse of a value set by the manufacturer is produced when it is triggered by a direct clear (DC) reset signal. R_{int} must be connected to V_{CC} and C_{ext}/R_{ext}, left open. Used in this way, the 74121 has an output pulse width of approximately 30–35 ns.

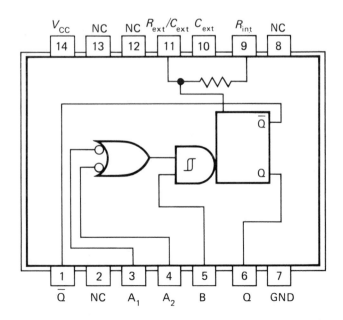

FIGURE 6–8

74121 IC Monostable Multivibrator

Inputs			Outputs	
A_1	A_2	B	Q	$\overline{Q}$
L	X	H	L	H
X	L	H	L	H
X	X	L	L	H
H	H	X	L	H
H	↓	H	⊓	⊔
↓	H	H	⊓	⊔
↓	↓	H	⊓	⊔
L	X	↑	⊓	⊔
X	L	↑	⊓	⊔

FIGURE 6–9

74121 IC Monostable Multivibrator Truth Table

Figure 6–9 shows the truth table for a 74121 monostable multivibrator. Notice that three inputs can trigger this IC one-shot. Inputs A_1 and A_2 trigger the one-shot if input B is held high. Inputs A_1 and A_2 trigger the one-shot on the trailing edge. When A_1 or A_2 switches from high to low, the one-shot generates an output pulse. Complementary output pulses appear at the Q and $\overline{Q}$ outputs. The duration of the output pulse is a function of the values of the external *RC* components. The manufacturer provides guidelines for selecting these values and charts for computing the pulse width for given values of *R* and *C*. See Figure 6–10.

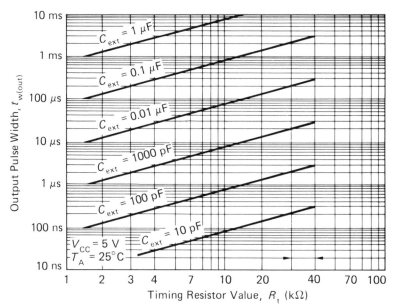

A. Output Pulse Width versus Timing Resistor Value

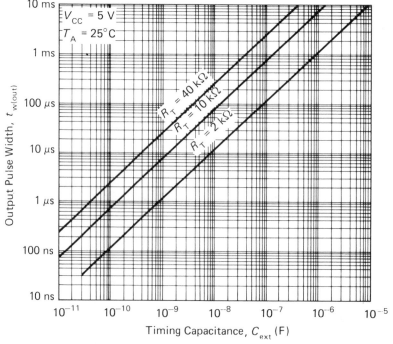

FIGURE 6–10

***R* and *C* Pulse Width Charts**

B. Output Pulse Width versus External Capacitance

The formula used for calculating the values of R and C is as follows:

$$t_{\text{w}} = 0.7(C_{\text{ext}} \times R_{\text{t}})$$

The charts reflect the result of these calculations. Usually, the value of external R is limited to approximately 50 kΩ, while any value of C from 10 pF to 100 μF can be used.

The only limitation for this IC one-shot multivibrator is that the output pulse width should not be adjusted to more than 90% of the input pulse frequency. The circuit must have time to reset. This limitation may be easier to understand if we study the two examples in Figure 6–11, which illustrate that there is practically no lower limit.

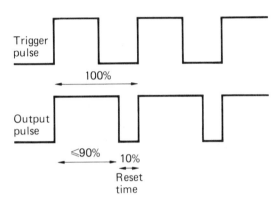

A. Example 1

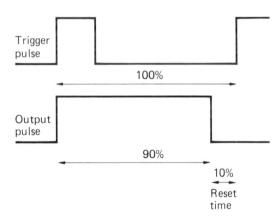

FIGURE 6–11

IC One-Shot Trigger and Output Pulses

B. Example 2

If inputs A_1 and A_2 are not used (held low) to trigger the IC one-shot, it can be triggered when input B switches from low to high. That is, input B triggers the one-shot on the leading edge of the input. This input is used primarily for the inhibiting or enabling of inputs A_1 and A_2. Note that some one-shots have a reset input similar to the asynchronous direct clear input of a JK flip-flop. Bringing this input low automatically terminates the output pulse during a timing period. When the one-shot is not triggered, the Q output is binary zero while the $\overline{Q}$ output is binary one. When the one-shot receives a trigger pulse, Q goes to binary one and $\overline{Q}$ goes to binary zero. If a reset pulse is applied during the timing period, the Q output will switch to binary zero, which immediately terminates the timing sequence.

The timing diagram shown in Figure 6–12 illustrates the operation of the IC one-shot. Input pulses P_1 and P_2 trigger the circuit into operation on the trailing edge. The output pulse, which has a duration of t, is defined by the values of R and C. Note that the timing interval terminates prior to the application of each new input pulse. On input pulse P_3, the one-shot is triggered, but the timing interval is cut short because of the occurrence of a reset pulse.

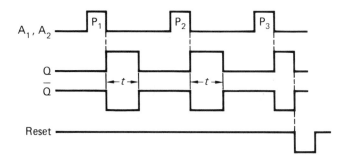

FIGURE 6–12

IC One-Shot Timing Diagram

Retriggerable One-Shot Multivibrator.   Another type of IC one-shot that is available to the digital designer is the *retriggerable one-shot multivibrator*. Most one-shots require a finite period of time in order to recover from a trigger pulse. Once a one-shot has been triggered and times out, only a short period of time is required for the capacitor to become recharged through the circuitry resistances. It is this recovery time that limits the upper duty cycle of most one-shots to approximately 90%. The retriggerable one-shot eliminates this problem. Its recovery time is nearly instantaneous, making almost 100% duty cycle outputs a possibility. A 100% duty cycle represents a constant binary one at output Q.

One of the advantages of the retriggerable one-shot multivibrator is its ability to generate output pulses of very long duration. When the values of the external RC components are adjusted to provide an output pulse duration that is longer than the interval between the input trigger pulses, the retriggerable one-shot remains in the triggered state for a substantial period of time. The input and output waveforms in the timing diagram of Figure 6–13 illustrate this effect.

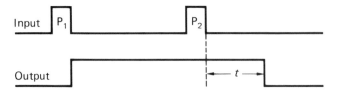

FIGURE 6–13

Retriggerable One-Shot
Multivibrator Timing Diagram

As shown in Figure 6–13, initially, the one-shot is in its normal stable state. It is triggered when the trailing edge of input pulse P_1 occurs. However, before it can complete its output pulse, which has a duration t as defined by the values of the external RC components, input pulse P_2 occurs. At the trailing edge of P_2, the first timing interval is automatically terminated, and a new timing interval is initiated so quickly that the output remains high. Note that another input pulse does not occur after input pulse P_2, and, therefore, the one-shot is allowed to time out. It generates its normal output pulse time (the RC time).

In addition to generating very long output pulses, the retriggerable one-shot multivibrator can be used as a missing pulse detector. A *missing pulse*, or *burst*, *detector circuit* indicates the start and end of a train of pulses. In this circuit, the timing cycle is continuously reset by the pulse train, which causes the output to stay high. A change in frequency, a missing pulse, or the end of a pulse train allows the timer to time out. The output then goes low, thereby indicating the missing pulse. The time delay is set to be longer than the normal time between pulses. (A typical detector with a 555 timer will be described later in the chapter.)

SELF-TEST EXERCISE 6–3

1. Define the term *monostable multivibrator*.
2. _____ the capacitance in a monostable multivibrator increases the pulse width of the output.
3. What value of R is needed for a $t_w = 10$ μs if a 1000 pF capacitor is used?
4. Most one-shots are limited to a 90% _____.
5. A retriggerable monostable multivibrator can be used as a _____.

 Do Experiment 6–2

555 Timer

The *555 timer* is probably the most versatile and popular timing circuit device on the market today, and the least expensive. This timer can be used in the astable or monostable mode of operation. It is a highly stable and flexible controller, capable of producing accurate time delays or oscillations. In the time delay mode of operation, the time is precisely controlled by one external resistor and one capacitor. In the astable mode of operation, the duty cycle and free-running frequency are accurately controlled by two external resistors and one capacitor.

The 555 timer has trailing-edge trigger and reset provisions. The output can source or sink to 200 mA (milliamperes). The timer is capable of producing time periods from microseconds to hours or days, depending on the external *RC* components. The supply voltage is 4.5 V to 18 V. The output pulse is TTL compatible. The width of the trigger pulse has no relationship to the output pulse.

555 Monostable Operation. For monostable operation, pins 6 and 7 are tied together, and the external capacitor is connected between this point and ground. See Figure 6–14. Initially, this external capacitor is held low (discharged) by a transistor inside the chip. The internal flip-flop is set upon application of a high-to-low trigger pulse to pin 2, which releases the short circuit across the capacitor and drives the output high for a period of time: $t = 1.1(R_aC)$. After this time, the internal comparator resets the flip-flop, which, in turn, discharges the capacitor and drives the output low.

Additional trigger pulses will not affect the circuit during the timing cycle when the output is high. Sending a high-to-low transition pulse to pin 4 (reset) will immediately bring the output low. It will remain low until another trigger pulse is applied.

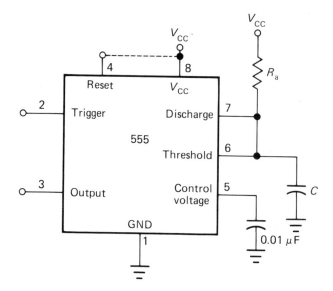

FIGURE 6–14

555 Monostable Timer

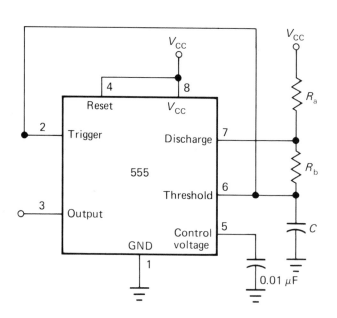

FIGURE 6–15

555 Astable Timer

555 Astable Operation.   For astable operation, pins 2 and 6 are tied together to allow the timer to trigger itself, forming an astable multivibrator. See Figure 6–15. Initially, the external capacitor charges through R_a and R_b and discharges through R_b. The duty cycle is set precisely by the ratio of these two resistors. The *charge time* (the time the output is high) is t_1; the *discharge time* (the time the circuit is low) is t_2. The total pulse time is T. These values are calculated as follows:

$$t_1 = 0.693(R_a + R_b)(C)$$

$$t_2 = 0.693(R_b)(C)$$

$$T = t_1 + t_2 = 0.693(R_a + 2R_b)(C)$$

555 Timer Applications.   The 555 timer has many applications. A typical application is shown in Figure 6–16. Here, the timer is used as a missing pulse detector. The input pulse train is connected to pin 2 of this

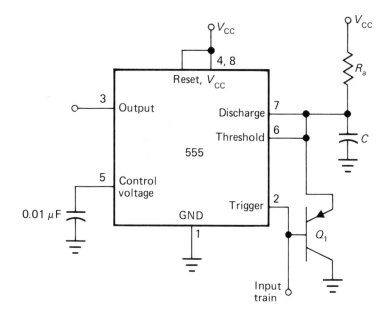

FIGURE 6–16

555 Timer Connected as a Missing Pulse Detector

circuit. The values of R_a and C are selected to give a normal output pulse that is slightly longer than the pulses being monitored. The output (pin 3) will go low only if a pulse is missing at pin 2. This output, however, will go high again when the next pulse is received.

If a circuit is needed that can remember the missing pulse, the output (pin 3) can be connected to an $\overline{RS}$ latch with an indicator, as shown in Figure 6–17. Any missing pulse will trigger the latch, and L_1 will light and stay lit until manually reset, regardless of additional missed pulses. If the number of missing pulses must be recorded, the output (pin 3) can be connected to a counter circuit. The possible uses for the 555 timer are many and depend on the particular application.

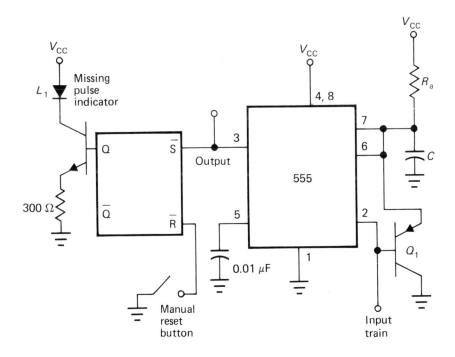

FIGURE 6–17

555 Timer Connected as a Missing Pulse Detector with Indicator

TRIGGERING OF EDGE-TRIGGERED FLIP-FLOPS

Figure 6–18 shows the necessary setup and hold time parameters that are characteristics of all clocked flip-flops including RS, JK, and D flip-flops. *Setup time* (t_S) is the amount of time that the data input must be held stable before the triggering edge of the clock occurs. If the flip-flop input is not held stable for the setup time or longer, the flip-flop may not trigger reliably. *Hold time* (t_H) is the amount of time that the data input must be held stable after the triggering edge of the clock has occurred. The flip-flop should not be retriggered during this time.

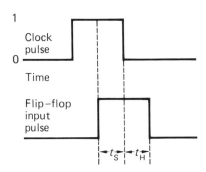

FIGURE 6–18

Setup Time and Hold Time

Remember, t_S is the amount of time that the data input must be available *before* the trigger pulse, and t_H is the amount of time that the data input must be available *after* the trigger pulse. Typically, setup and hold times are in the nanosecond range, with t_S = 5 to 50 ns and $t_H < 10$ ns. If $t_H = 0$, the output change occurs at the same time as the input clock transition.

TWO-PHASE (DELAYED) CLOCK CIRCUITS

Many microprocessor-based circuits and MOS circuits need two-phase clock pulses for their timing. *Two-phase*, or *delayed*, *clock circuits* are also referred to as *nonoverlapping clocks*.

There are numerous ways to generate a two-phase clock. One typical two-phase clock circuit is shown in Figure 6–19. Both clocks are generated at half the input clock frequency by using a JK flip-flop and two NOR gates. Some circuitry may be added to shift the logic levels of the signals required. The delayed pulses occur midway between the clock pulses, as illustrated by C_{ϕ_1} and C_{ϕ_2} in the timing diagram of Figure 6–20.

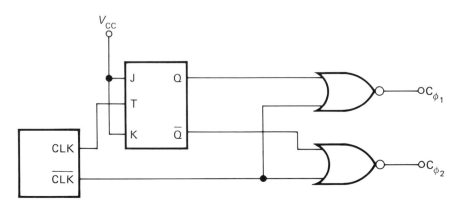

FIGURE 6–19

Two-Phase Clock Circuit

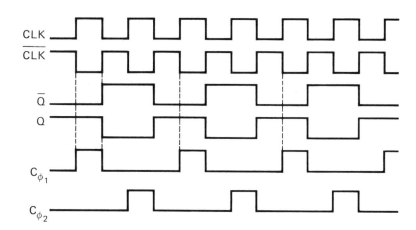

FIGURE 6-20

Two-Phase Clock Timing
Diagram

SELF-TEST EXERCISE 6-4

1. The 555 timer can be used only as an astable multivibrator. True or False?
2. All clocked flip-flops have setup and hold time parameters. True or False?
3. If an input is not held stable, the flip-flop (will, will not) trigger reliably.
4. Typically, t_S and t_H are in the _____ range.
5. _____ separate clock pulses so that reliable triggering can be achieved.
6. If _____ = 0, the output change occurs at the same time as the input clock transition.

 STOP Do Experiment 6-3

SUMMARY

The master clock is used by a complete digital system as a reference. The master clock tells all parts of the system when and how fast work will be done. Subordinate clocks are derived from the master clock and operate at a frequency less than, but synchronized to, the master clock.

Clocks operated on a frequency determined by resistor–capacitor (RC) networks or, for precision timing, by crystal-controlled oscillators. The maximum frequency for a clock is determined by the slowest propagation delay in the system under control. That is, the clock cannot tell the other parts of the system to do work faster than they can do it.

An astable (free-running) multivibrator circuit has two momentarily stable states. It continuously switches from one state to the other. Either an RC network or a crystal determines how fast it alternates between states.

A monostable (one-shot) multivibrator circuit produces an output pulse each time it is triggered. It is not free running. The duration of the output pulse is determined, not by the trigger signal, but by the RC network.

A missing pulse (burst) detector is a type of monostable circuit. It must be continually triggered; otherwise, the output goes low. This low output can then be used to signal a malfunction. For example, this low output can be used to set an RS flip-flop whose output lights an LED. The RS flip-flop will stay set (light on) until manually reset, even though the pulse detector goes high.

The duty cycle of a circuit is the amount of time that the circuit is high compared to the amount of time that it is low. A 90% duty cycle means that it is "on duty" (high) 90% of the time and "off duty" (low) 10% of the time. (If we were to work a 90% duty cycle, we would be at work 21.6 hours out of every 24 hours!)

The 555 timer is probably the most versatile and popular integrated circuit on the market today. It can be used as an astable or monostable circuit. It operates on a wide voltage range and can easily be interfaced to TTL circuits. And, although its applications are many, its price is very low.

Setup time of an IC is the amount of time that the data input must be held stable before the circuit is triggered. Hold time is the amount of time that the data input must be held stable after the circuit is triggered. The propagation delay is the length of time that the circuit takes to respond to the data input.

Two-phase (delayed) clocks are used for many applications because there is no overlap between the leading or trailing edges of the clock pulse. The circuits, therefore, have time to set up for the next signal.

CHAPTER 6
REVIEW EXERCISES

1. What is the function of the master clock?

2. Draw the waveform for a 10 MHz clock. Label the leading and the trailing edges. What is the time period for this clock? Label it.

3. The proper type of oscillator for precision timing is:
   a. monostable
   b. astable
   c. *RC* network
   d. crystal

4. The subordinate clock is derived from _____.

5. Define the term *astable multivibrator*.

6. Define the term *monostable multivibrator*.

7. What is the circuit parameter that determines the maximum operating clock frequency selected for a digital circuit?

8. Draw the waveforms for a two-phase clock.

9. If the capacitance is decreased in an *RC* network, the output frequency is (increased, decreased).

10. If the capacitance is increased in an *RC* network, the time period is (increased, decreased).

11. If the resistance is increased in an *RC* network, the frequency of the output is (increased, decreased).

12. If the resistance is increased in an *RC* network, the time period is (increased, decreased).

13. What is the purpose of an IC buffer gate in an oscillator circuit?

14. What is the most popular timing circuit device as cited in this chapter? Why?

15. What is setup time?

16. What is hold time?

17. Briefly describe how a missing pulse detector works.

EXPERIMENT 6–1 | Astable Clock Circuits

PURPOSE

This experiment is designed to show the operation of astable clock circuits. The first circuit is designed to oscillate at the parameters set by the RC components. The second circuit is crystal controlled and oscillates at the crystal frequency. You will build these clocks from integrated circuits.

EQUIPMENT

—2 1000 Ω resistors (R_1, R_2)
—2 510 Ω resistors (R_3, R_4)
—1 3.57 MHz crystal (XTAL)
—2 1 µF capacitors (C_1, C_2)
—1 270 pF capacitor (C_3)
—1 SN74LS04 IC inverter (U_1)
—1 digital experimenter (Equip$_1$)
—1 10 MHz or better oscilloscope (Equip$_2$)

PROCEDURE: PART I

Step 1 Construct the circuit as shown in Figure 6E1–1.

NOTE: Clock circuits are especially sensitive to capacitance and resistance values; therefore, the way in which you wire up this circuit will affect the output. Take care in using wire leads as short as possible. Each piece of wire has a resistance and capacitance value as does the breadboard; these values can affect very critical timing circuits. The circuits built in this experiment are less susceptible to this problem, but for best results use good breadboarding techniques.

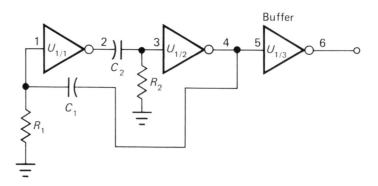

FIGURE 6E1–1

Step 2 Connect the oscilloscope lead to the output of the buffer gate.

ACTIVITY

Measure the frequency of this clock circuit. Record the value.

NOTE: The frequency should be approximately 450 Hz (1/2.2RC). If the circuit is not oscillating, check to be sure you have connected V_{CC} and ground to the IC.

ACTIVITY

Draw the waveform you see on the oscilloscope. It should look like the waveform in Figure 6E1–2.

FIGURE 6E1–2

ACTIVITY

Look at pins 1 and 3 with the oscilloscope. You should see a charge/discharge pattern like the one shown in Figure 6E1–3.

FIGURE 6E1–3

PROCEDURE: PART II

Step 1 Modify the circuit as shown in Figure 6E1–4. Again, wire carefully.

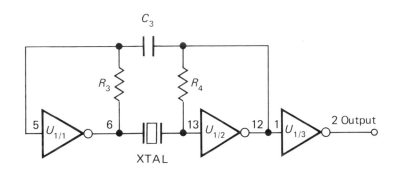

FIGURE 6E1–4

Step 2 Connect the oscilloscope lead to the output of the buffer gate.

ACTIVITY

Measure the frequency of this clock circuit. Record the value.

NOTE: The frequency should be approximately 3.57 MHz. If the circuit is not oscillating, check the wiring.

ACTIVITY

Draw the waveform you see on the oscilloscope. The better the scope you use, the squarer the waveform you will see. Figure 6E1–5 shows the output on a 10 MHz scope; Figure 6E1–6 shows the output on a 100 MHz scope.

NOTE: The troubleshooter should realize that the response of the oscilloscope may be the problem, not the circuit.

FIGURE 6E1–5

FIGURE 6E1–6

| EXPERIMENT 6–2 | # Monostable Multivibrator Circuits |

PURPOSE

This experiment is designed to show the operation of a monostable multivibrator circuit. You will build a one-shot multivibrator with an SN74121 integrated circuit. You will then examine the relationship of capacitor value to pulse width.

EQUIPMENT

—1 150 kΩ resistor (R_1)
—1 10 μF capacitor (C_1)
—1 100 μF capacitor (C_2)
—1 0.01 μF capacitor (C_3)
—1 1000 μF capacitor (C_4)
—1 SN74121 monostable multivibrator IC (U_1)
—1 digital experimenter (Equip$_1$)
—1 10 MHz or better oscilloscope (Equip$_2$)

PROCEDURE

Step 1 Construct the circuit as shown in Figure 6E2–1.

NOTE: Input A_1 is connected to logic switch $\overline{A}$, and input A_2 is connected to logic switch $\overline{B}$. Input B is connected to data switch SW_1. SW_1 should be high to enable A_1 and A_2.

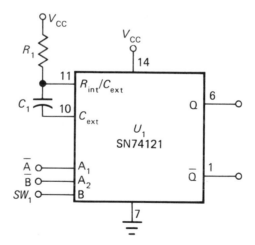

FIGURE 6E2–1

Step 2 Connect the Q output to logic indicator L_1 and to the oscilloscope input.

Step 3 Apply power.

ACTIVITY

Complete the truth table in Figure 6E2–2.

NOTE: You should see an output pulse width equal to approximately 1 second (PW = 0.7RC).

157

FIGURE 6E2–2

A₁	A₂	B	Q	Q̄
H	↓	H		
↓	H	H		
↓	↓	H		

Step 4 Turn off the power.

Step 5 Move input A_2 from logic switch $\overline{B}$ to logic switch A.

Step 6 Move input B from data switch SW_1 to logic indicator B.

Step 7 Turn on the power.

ACTIVITY

Complete the truth table in Figure 6E2–3.

NOTE: You should see the same output pulse that you saw earlier.

FIGURE 6E2–3

A₁	A₂	B	Q	Q̄
H	L	↑		
L	H	↑		

Step 8 Replace C_1 with C_2.

Step 9 Trigger the monostable multivibrator.

NOTE: The pulse width for this circuit should be approximately 10 seconds. If your circuit does not time out at exactly 10 seconds, remember, the tolerances of the resistor and capacitor will affect the time. If you use precision resistors and capacitors, you will get excellent results.

Step 10 Replace C_2 with C_4.

NOTE: The pulse width should now be approximately 100 seconds. Very large capacitors produce long pulse widths; very small capacitors produce very fast pulse widths.

Step 11 Connect input B to a 1 Hz clock.

Step 12 Replace C_4 with C_3.

ACTIVITY

Look at Q with the oscilloscope. You will see a 1 ms pulse every second. Figure 6E2–4 shows a timing diagram for this circuit.

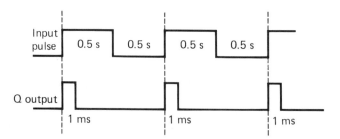

FIGURE 6E2–4

EXPERIMENT 6–3 | 555 Timer Circuits

PURPOSE

This experiment is designed to show the operation of 555 timer circuits. The first circuit that you will build is a 555 monostable multivibrator. You will compare the actual output pulse of this circuit to the calculated pulse width. The second circuit that you will build is a 555 missing pulse detector.

EQUIPMENT

—1 68 kΩ resistor (R_1)
—1 100 μF capacitor (C_1)
—1 1000 μF capacitor (C_2)
—1 0.01 μF capacitor (C_3)
—1 10 μF capacitor (C_4)
—1 555 timer IC (U_1)
—1 SN7400 NAND gate IC (U_2)
—1 2N3906 PNP transistor (Q_1)
—1 normally closed (NC) push button (PB$_1$)
—1 digital experimenter (Equip$_1$)

PROCEDURE: PART I

Step 1 Construct the circuit as shown in Figure 6E3–1.

NOTE: The 555 timer triggers on the low-to-high transition of logic switch $\overline{A}$ at pin 2.

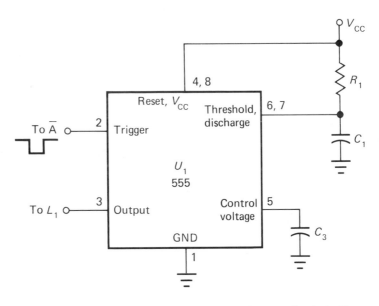

FIGURE 6E3–1

Step 2 Connect the output (pin 3) of the 555 timer to logic indicator L_1.

Step 3 Turn on the power.

Step 4 Depress logic switch $\overline{A}$ to simultaneously start the timing.

NOTE: This circuit starts timing immediately.

ACTIVITY

Measure how long the output pulse is high. Compare this measured time delay with the calculated time ($t = 1.1RC$).

> **Check:** Is the calculated time different from the actual measured time? If so, why?

Step 5 Turn off the power to replace C_1 with C_2.

ACTIVITY

Calculate the expected time pulse.

Step 6 Turn on the power and trigger the circuit.

ACTIVITY

Measure the pulse.

> **Check:** Does the time agree with the calculated time? If not, why?

NOTE: The output pulse width of the 555 timer is not dependent on the input pulse width; therefore, it is used as a time delay or "pulse stretcher."

PROCEDURE:
PART II

Step 1 Construct the circuit as shown in Figure 6E3–2.

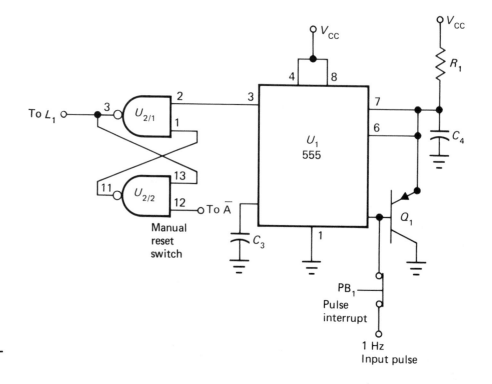

FIGURE 6E3–2

NOTE: This circuit is a missing pulse detector. When a missing pulse is detected, logic indicator L_1 will light and stay lit until it is reset by logic switch $\overline{A}$. The normally closed push button PB_1 is used to temporarily interrupt the input pulses. You may wish to connect the input pulse train to logic indicator L_2, and the output (pin 3) of the 555 timer to L_3, for a better view of what is going on in the circuit.

Step 2 Turn on the power.

NOTE: L_1 should be off. If it is not, reset the circuit with logic switch $\overline{A}$.

Step 3 Interrupt the pulse train by pressing PB_1.

NOTE: L_1 should now light and stay lit.

Step 4 Reset the latch with logic switch $\overline{A}$.

Step 5 Interrupt the pulse train again.

NOTE: L_1 will again light. Since L_1 stays on after the pulse train is no longer interrupted, this circuit is useful only as an alarm or warning that the input train had an error. It does not say when an error occurred or how many errors occurred. More circuitry is needed if the number of errors is to be recorded.

Encoders, Decoders, Multiplexers, and Demultiplexers

OBJECTIVES

After studying this chapter, you will be able to:

1. Describe the operation of encoding, priority encoding, decoding, multiplexing, and demultiplexing circuits.
2. List typical applications for these circuits.
3. Using AND and inverter gates, draw a logic gate decoder for a given binary or BCD number.
4. Define a one-to-N decoder (N is a given decimal value).
5. Given a schematic, build a decoding circuit for a 7-segment display.

INTRODUCTION

Digital circuits often need "interpreters" to communicate with the outside world. Encoders and decoders function as interpreters of data. Digital circuits also need "traffic directors." Multiplexers and demultiplexers are used to route data. This chapter focuses on encoding, decoding, multiplexing, and demultiplexing circuits and describes how these combinational circuits perform operations on input data.

ENCODERS

An *encoder* is an electronic circuit that is used to convert a single selected input into a binary equivalent as the output. It, therefore, generates a specific binary code. An encoder is a link between computer personnel and computer. That is, it takes a decimal value input and converts it to a binary output that can be understood by a digital circuit or a computer.

MSI Encoders

Two types of TTL MSI encoders available are—the 74148 8-line to 3-line encoder and the 74147 10-line to 4-line encoder. An *8-line to 3-line encoder* takes an 8-line input equivalent to decimal values 0 through 7 and converts it to the 3-bit binary equivalent (000_2 to 111_2). Logic lows (active lows) are used as the input conditions. The outputs are also active lows. Similarly, a *10-line to 4-line encoder* takes the 10-line decimal equivalent to 0 through 9 and encodes it to a 4-bit BCD output.

A typical application of an encoder is the translation of a decimal keyboard input into a binary output code. For each key pressed, the 3- or 4-bit binary code corresponding to the number keyed in is generated. Study, for example, the truth table for the 74148 zero-to-seven encoder shown in Figure 7–1 and the schematic for the decimal keyboard circuit shown in Figure 7–2.

Inputs									Outputs				
EI	0	1	2	3	4	5	6	7	A_2	A_1	A_0	GS	EO
H	X	X	X	X	X	X	X	X	H	H	H	H	H
L	H	H	H	H	H	H	H	H	H	H	H	H	L
L	X	X	X	X	X	X	X	L	L	L	L	L	H
L	X	X	X	X	X	X	L	H	L	L	H	L	H
L	X	X	X	X	X	L	H	H	L	H	L	L	H
L	X	X	X	X	L	H	H	H	L	H	H	L	H
L	X	X	X	L	H	H	H	H	H	L	L	L	H
L	X	X	L	H	H	H	H	H	H	L	H	L	H
L	X	L	H	H	H	H	H	H	H	H	L	L	H
L	L	H	H	H	H	H	H	H	H	H	H	L	H

H = high level (same as 1)
L = low level (same as 0)
X = irrelevant

FIGURE 7–1

74148 Truth Table

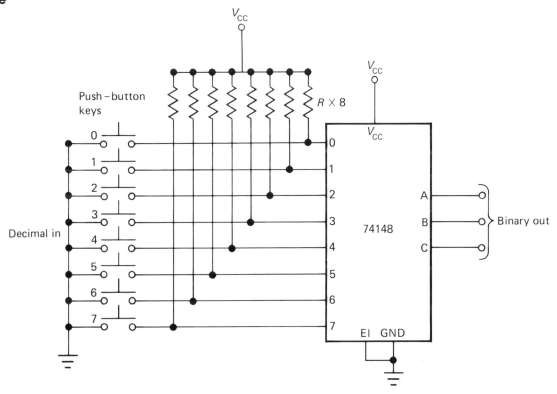

FIGURE 7–2

Decimal Keyboard Circuit

We can see from Figures 7–1 and 7–2 that, when the "3" key is pressed, the binary output 100_2 is generated; when the "7" key is pressed, the binary output 000_2 is generated. This conversion may be confusing to you at first because you are accustomed to seeing a binary 3 as 011_2. But, look at the pin assignment diagram in Figure 7–3. The small circles at the A, B, and C outputs indicate that the generated signal is inverted before it is outputted. Therefore, 011_2 is inverted to 100_2 before it is sent out. Similarly, the inputs are inverted so that active lows are used as the input conditions.

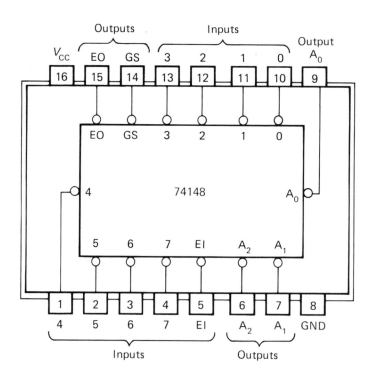

FIGURE 7–3

74148 Pin Assignments

Look at Figure 7–3 again. Note that EI is the enabled inputs control pin for this IC, EO is the enabled outputs indicator, and GS is a group signal output. When the GS output goes high, it indicates that no data inputs are low. A low on this pin indicates that the encoder is enabled and has an active low input signal.

Priority Encoders

The 74147 and the 74148 have additional circuitry that allows the largest activated digit to be given priority over other inputs. That is, if two or more inputs are activated at the same time, the output will react only to the largest digit. For example, if the 3, 5, and 7 inputs of 74148 were activated at the same time, the output would be 1000_2, the active low logic BCD equivalent of a decimal 7. The 3 and 5 would be ignored. Because the largest digit is given priority over the other inputs, the 74147 and the 74148 are known as *priority encoders*.

SELF-TEST EXERCISE 7–1

1. An _____ generates a particular BCD or binary output word for a single selected input.
2. An encoder can be used to interface a _____ to a digital circuit.
3. If the 6 input of a 74148 encoder were activated, the output would be _____. (Remember, the circuit works on active low logic.)
4. If the 2, 4, and 9 inputs of a 74147 encoder were activated at the same time, the output code would be _____.
5. Priority encoders have additional circuitry that allows the circuit to ignore all but the (largest, smallest) activated input.

STOP   Do Experiment 7–1

DECODERS

A *decoder*, the opposite of an encoder, converts binary input data into a single output signal. A decoder is a link between computer and computer personnel. That is, it translates binary numbers from a digital circuit or a computer into decimal equivalent outputs. Some special decoders have more than one output for each binary input. These special decoders are used to operate (drive) 7-segment displays.

Basic and MSI Decoders

As shown in Figure 7–4, a basic decoder can be constructed from an AND gate. The circuit in Figure 7–4 detects the presence of a binary 11_2. Because it detects (goes high on) only one of the four possible binary inputs, the circuit is known as a *one-of-four decoder*.

The circuit in Figure 7–4 can be made to detect other binary numbers by attaching various logic circuits. For example, to detect 10_2, the circuit shown in Figure 7–5 would be used. This circuit is still a one-of-four decoder, however, since it detects only one of the four possible input states.

FIGURE 7–4

Detect 11_2 Decoder

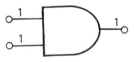

FIGURE 7–5

Detect 10_2 Decoder

Most circuits need to detect the presence of all combinations represented by the binary input words. Figure 7–6 shows a typical *one-to-four decoder*. Here, four AND gates are used to detect all of the four possible input states and decode them to the appropriate decimal equivalent output. The truth table for this decoder, shown in Figure 7–7, indicates that, at any given time, only one output line is enabled (is in the one state).

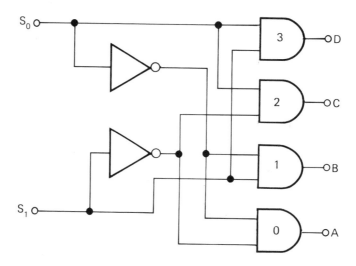

FIGURE 7–6

One-to-Four Decoder

Inputs		Outputs			
S_1	S_0	D	C	B	A
		1	2	1	0
0	0	L	L	L	H
0	1	L	L	H	L
1	0	L	H	L	L
1	1	H	L	L	L

FIGURE 7–7

One-to-Four Decoder Truth Table

A one-to-four decoder can be considered as a *binary-to-decimal converter*. It converts a binary number into an output that represents the decimal numbers 0 through 3. Similarly, BCD-to-decimal decoders, like the one shown in Figure 7–8, can be constructed.

In a *BCD-to-decimal decoder*, the input is a 4-bit binary word representing 0000_2 through 1001_2. Ten AND gates are used to decode these binary words to the decimal numbers 0 through 9. When a BCD code is applied to the input, one of the ten output lines goes high or low and indicates the presence of the BCD code. This decoding circuit can be used to turn on (or off) a peripheral device when a particular binary count is reached. The circuit does not recognize the 4-bit words that are not included in the 8–4–2–1 BCD code.

Today, the entire circuitry for many decoding operations is available in MSI ICs. Not having to construct decoders from individual gates saves considerable time, money, and space in a digital circuit. The 7442, for example, is an MSI decoder that decodes BCD to decimal (4-line to 10-line). Its truth table is shown in Figure 7–9; its pin assignments, in

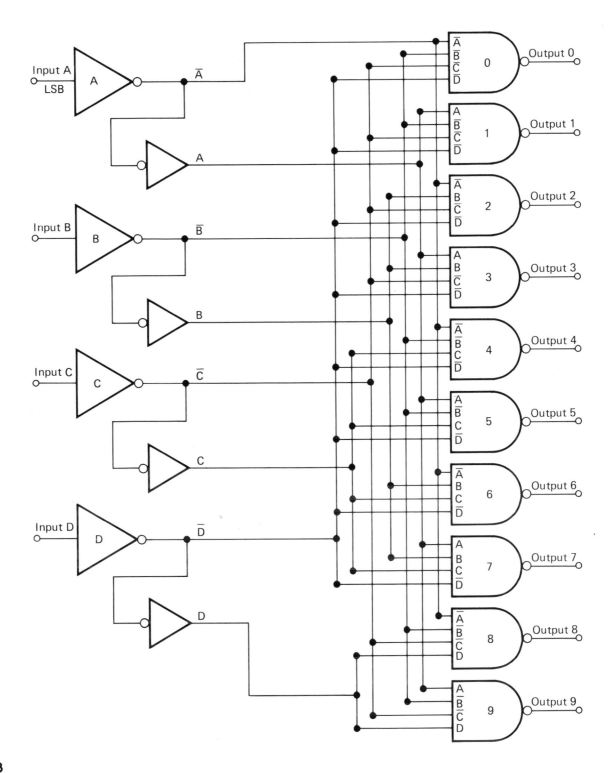

FIGURE 7–8

BCD-to-Decimal Decoder

Figure 7–10. Notice that the outputs of the 7442 are active low (go low when activated). All outputs remain inactive (high) for the invalid states, 1010_2 through 1111_2. Only one output can be active at a time if the chip is to operate correctly.

No.	BCD inputs				Decimal outputs									
	D	C	B	A	0	1	2	3	4	5	6	7	8	9
0	L	L	L	L	L	H	H	H	H	H	H	H	H	H
1	L	L	L	H	H	L	H	H	H	H	H	H	H	H
2	L	L	H	L	H	H	L	H	H	H	H	H	H	H
3	L	L	H	H	H	H	H	L	H	H	H	H	H	H
4	L	H	L	L	H	H	H	H	L	H	H	H	H	H
5	L	H	L	H	H	H	H	H	H	L	H	H	H	H
6	L	H	H	L	H	H	H	H	H	H	L	H	H	H
7	L	H	H	H	H	H	H	H	H	H	H	L	H	H
8	H	L	L	L	H	H	H	H	H	H	H	H	L	H
9	H	L	L	H	H	H	H	H	H	H	H	H	H	L
Invalid states					No outputs									
10	H	L	H	L	H	H	H	H	H	H	H	H	H	H
11	H	L	H	H	H	H	H	H	H	H	H	H	H	H
12	H	H	L	L	H	H	H	H	H	H	H	H	H	H
13	H	H	L	H	H	H	H	H	H	H	H	H	H	H
14	H	H	H	L	H	H	H	H	H	H	H	H	H	H
15	H	H	H	H	H	H	H	H	H	H	H	H	H	H

FIGURE 7–9

Truth Table for 7442 BCD-to-Decimal Decoder

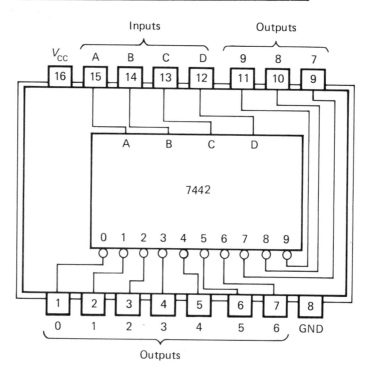

FIGURE 7–10

Pin Assignments for 7442 BCD-to-Decimal Decoder

Octal and Hexadecimal Decoders

Two other types of decoders are widely used. They are the *one-of-eight*, or *octal, decoder* and the *one-of-sixteen*, or *hexadecimal, decoder*. The octal decoder uses only three inputs. A BCD-to-decimal decoder can be used as an octal decoder with the fourth input tied to zero and the eighth and ninth outputs ignored. A hexadecimal decoder is a 4-bit binary decoder that decodes all sixteen states.

The 74138 is an MSI 3-line to 8-line, one-of-eight, octal decoder. Its pin assignments and truth table are shown in Figure 7–11. Notice

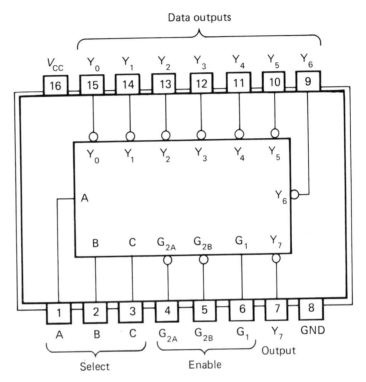

A. Pin Assignments

Inputs					Outputs							
Enable		Select										
G_1	G_2	C	B	A	Y_0	Y_1	Y_2	Y_3	Y_4	Y_5	Y_6	Y_7
X	H	X	X	X	H	H	H	H	H	H	H	H
L	X	X	X	X	H	H	H	H	H	H	H	H
H	L	L	L	L	L	H	H	H	H	H	H	H
H	L	L	L	H	H	L	H	H	H	H	H	H
H	L	L	H	L	H	H	L	H	H	H	H	H
H	L	L	H	H	H	H	H	L	H	H	H	H
H	L	H	L	L	H	H	H	H	L	H	H	H
H	L	H	L	H	H	H	H	H	H	L	H	H
H	L	H	H	L	H	H	H	H	H	H	L	H
H	L	H	H	H	H	H	H	H	H	H	H	L

$G_2 = G_{2A} + G_{2B}$
H = high level
L = low level
X = irrelevant

FIGURE 7–11

74138 Octal Decoder

B. Truth Table

that the G_1 enable must be high, while the G_{2A} and G_{2B} enables are low for the decoding operation. This circuit is another active low output circuit.

The 74154 is a 4-line to 16-line, one-of-sixteen, hexadecimal decoder. Its pin assignments and truth table are shown in Figure 7–12.

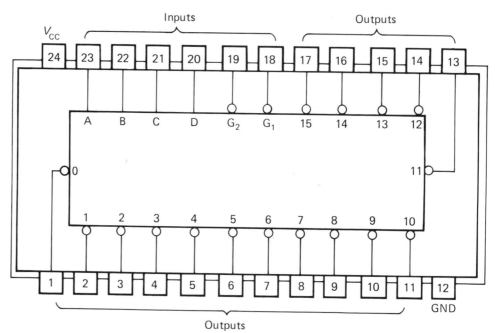

A. Pin Assignments

Inputs						Outputs															
G_1	G_2	D	C	B	A	0	1	2	3	4	5	6	7	8	9	10	11	12	13	14	15
L	L	L	L	L	L	L	H	H	H	H	H	H	H	H	H	H	H	H	H	H	H
L	L	L	L	L	H	H	L	H	H	H	H	H	H	H	H	H	H	H	H	H	H
L	L	L	L	H	L	H	H	L	H	H	H	H	H	H	H	H	H	H	H	H	H
L	L	L	L	H	H	H	H	H	L	H	H	H	H	H	H	H	H	H	H	H	H
L	L	L	H	L	L	H	H	H	H	L	H	H	H	H	H	H	H	H	H	H	H
L	L	L	H	L	H	H	H	H	H	H	L	H	H	H	H	H	H	H	H	H	H
L	L	L	H	H	L	H	H	H	H	H	H	L	H	H	H	H	H	H	H	H	H
L	L	L	H	H	H	H	H	H	H	H	H	H	L	H	H	H	H	H	H	H	H
L	L	H	L	L	L	H	H	H	H	H	H	H	H	L	H	H	H	H	H	H	H
L	L	H	L	L	H	H	H	H	H	H	H	H	H	H	L	H	H	H	H	H	H
L	L	H	L	H	L	H	H	H	H	H	H	H	H	H	H	L	H	H	H	H	H
L	L	H	L	H	H	H	H	H	H	H	H	H	H	H	H	H	L	H	H	H	H
L	L	H	H	L	L	H	H	H	H	H	H	H	H	H	H	H	H	L	H	H	H
L	L	H	H	L	H	H	H	H	H	H	H	H	H	H	H	H	H	H	L	H	H
L	L	H	H	H	L	H	H	H	H	H	H	H	H	H	H	H	H	H	H	L	H
L	L	H	H	H	H	H	H	H	H	H	H	H	H	H	H	H	H	H	H	H	L
L	H	X	X	X	X	H	H	H	H	H	H	H	H	H	H	H	H	H	H	H	H
H	L	X	X	X	X	H	H	H	H	H	H	H	H	H	H	H	H	H	H	H	H
H	H	X	X	X	X	H	H	H	H	H	H	H	H	H	H	H	H	H	H	H	H

H = high level
L = low level
X = irrelevant

B. Truth Table

FIGURE 7–12

74154 Hexadecimal Decoder

Here, both G_1 and G_2 strobe inputs must be low for the circuit to operate as a decoder. The circuit can also operate as a demultiplexer. (Demultiplexers will be discussed later in this chapter.)

BCD to 7-Segment Decoders

A *BCD to 7-segment decoder* is a special kind of decoder that is used with 7-segment displays. For a 4-bit binary BCD input, it gives a 7-bit output code that is used to drive the displays.

A *7-segment display* is an electronic component that displays the numbers 0 through 9 by illumination of two or more of seven segments in a specially arranged pattern. That is, there are seven segments, and numbers are formed by lighting selected segments. A typical example is shown in Figure 7–13. This 7-segment display format is used in calculators, watches, clocks, and electronic equipment such as digital voltmeters and frequency counters.

Technology has advanced to the point where displays now come in different types, colors, and sizes to suit varied applications. Although light-emitting diodes (LEDs), liquid crystal displays (LCDs), and incandescent (filament), gas-flow discharge, and fluorescent displays use different voltages, the decoder used in each type is basically the same.

The most widely and easily used type of display is the *light-emitting diode* (LED), a transparent semiconductor diode that emits light in a 7-segment pattern. LEDs are small, rugged, and compatible with TTL devices such as the BCD to 7-segment decoder/driver.

The *fluorescent display* is blue-green in color. It is made of a material that fluoresces (glows) in the presence of low voltages.

The *liquid crystal display* (LCD) is made of a material that scatters light instead of generating it as do LED and fluorescent displays. LCDs use the least amount of power, which makes them useful for battery-operated devices. Unfortunately, they have a short operating life and are sensitive to excessive ambient heat.

The truth table for a BCD to 7-segment decoder/driver is shown in Figure 7–14. The inputs are D, C, B, and A (8–4–2–1). The outputs (the different segments that must be lit to display a particular decimal number) are a, b, c, d, e, f, and g.

Aside from producing the decimal numbers 0 through 9, 7-segment decoder/drivers have other capabilities. For example, for testing the seven segments to see whether they are operational, a lamp test ($\overline{\text{LT}}$) line is available. A low at the $\overline{\text{LT}}$ pin will cause all segments to glow, displaying an 8. Also, for suppressing the display of leading zeros, a blanking input ($\overline{\text{RBI}}$) line is available. This line is required when a group of displays is used to represent large decimal numbers. The zero blanking control allows, for example, the number 1679 to be displayed as 1679—not 001679—when six displays are used. Many decoder/drivers also have an overriding blanking input ($\overline{\text{BI}}$) line, which can be used either to control the intensity of the display by pulsing it or to inhibit all outputs. Pulsing the display can be done at any time by using this input since it does not inhibit the counting operation. A pulsing display is often used when a certain count indicates a dangerous condition in an industrial control application. The pulsing display is readily observed.

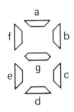

FIGURE 7–13

7-Segment Display Format

	Inputs					Outputs						
Display	Decimal value	D	C	B	A	a	b	c	d	e	f	g
0	0	0	0	0	0	1	1	1	1	1	1	0
1	1	0	0	0	1	0	1	1	0	0	0	0
2	2	0	0	1	0	1	1	0	1	1	0	1
3	3	0	0	1	1	1	1	1	1	0	0	1
4	4	0	1	0	0	0	1	1	0	0	1	1
5	5	0	1	0	1	1	0	1	1	0	1	1
6	6	0	1	1	0	1	0	1	1	1	1	1
7	7	0	1	1	1	1	1	1	0	0	0	0
8	8	1	0	0	0	1	1	1	1	1	1	1
9	9	1	0	0	1	1	1	1	0	0	1	1
⊏	10	1	0	1	0	0	0	0	1	1	0	1
⊐	11	1	0	1	1	0	0	1	1	0	0	1
⊔	12	1	1	0	0	0	1	0	0	0	1	1
⊆	13	1	1	0	1	1	0	0	1	0	1	1
t	14	1	1	1	0	0	0	0	1	1	1	1
Blank	15	1	1	1	1	0	0	0	0	0	0	0

Decimal number not displayed (values 10–15)

FIGURE 7–14

Truth Table for BCD to 7-Segment Decoder/Driver

Since decoder ICs are also drivers, they contain a transistor circuit that drives the LED segment. Figure 7–15 shows a typical decoder/driver. The open collector saturated transistor is used as the output to the LED segment. When the transistor conducts, the output goes low and current flows through the segment, turning it on.

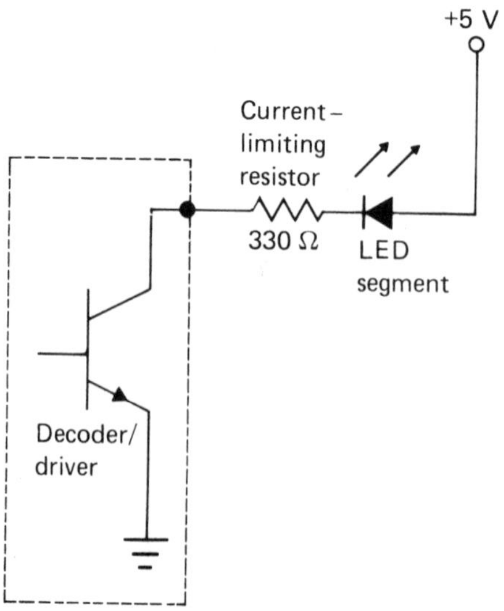

FIGURE 7–15

Decoder/Driver Circuit

Current-limiting resistors are required with LEDs to limit the current flow and thus prevent destructive currents. Each segment in a

7-segment device must have a separate resistor to prevent unequal segment brightness. Some special 7-segment decoder/drivers have built-in current sources, which make the resistors unnecessary. Note that a direct V_{CC} or ground should never be applied to a 7-segment display because the segment will be irreparably damaged. A 330 Ω current-limiting resistor is usually sufficient to prevent device failure.

SELF-TEST EXERCISE 7–2

1. A decoder is used to detect the presence of a _____ word.
2. A one-of-eight decoder is (a hexadecimal, an octal) decoder.
3. The simplest decoder is an AND gate. True or False?
4. The decoder circuit in Figure 7–16 decodes only the number _____.

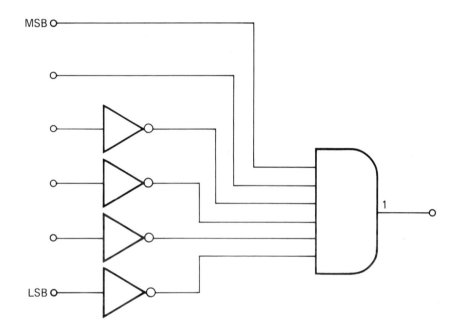

FIGURE 7–16 _____

5. A _____ is used at the input of most 7-segment displays to limit the current flow.
6. To display a decimal 2 with a 7-segment display, segments _____ must be lit.
7. A _____-to-decimal decoder has six invalid states.

STOP
Do Experiment 7–2

Do Experiment 7–3

MULTIPLEXERS

A *multiplexer*, or *data selector*, is an electronic circuit that is used to select and route any of two or more input signals to a single output. That is, it allows only one input at a time to get through to the output. It is the electronic equivalent of a multiposition switch. It has from two to sixteen data inputs and one output.

Analog as well as digital types are available. For analog applications, relays and bipolar or MOSFET switches are widely used. For digital applications, multiplexers either are constructed with logic gates or are available in MSI form.

The term *multiplex* refers to the handling of data over common lines. Data transfer is accomplished by time-sharing the use of these lines, or *buses*, as they are often called. Data transfers in many computer systems are accomplished by multiplexing circuits and by time-sharing techniques. The time sharing of bus lines is an economical way of transferring data. And, when bus lines can be used, they save considerable space on a printed circuit board, which allows more room for other components and functions.

Figure 7–17 shows a simple *2-input multiplexer*, its truth table, and its equivalent switching circuit. This multiplexing circuit has two input sources and one output. Either one of the two input sources may be selected and passed to the output. The routing of the desired input data to the output is controlled by the select input lines, or *address lines*. Similar to the knob on a multiposition switch, a select input decides which line or lines are tied to the common or output.

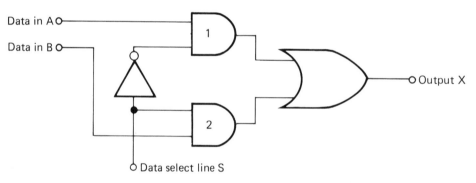

A. 2-Line to 1-Line Multiplexer

S	X
0	A
1	B

B. Truth Table

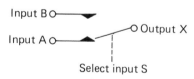

C. Equivalent Switching Circuit

FIGURE 7–17

Simple 2-Input Multiplexer

As illustrated in Figure 7–18, 2-input multiplexers can be combined to form two multibit words. These words can be output selectively: The D_1–C_1–B_1–A_1 word is output when the select line (S) is high; the D_2–C_2–B_2–A_2 word is output when S is low. The enable line (E) must be held low for the transmission of data to occur. When E is high, the output of all gates is inhibited, which produces a zero output at A, B, C, and D.

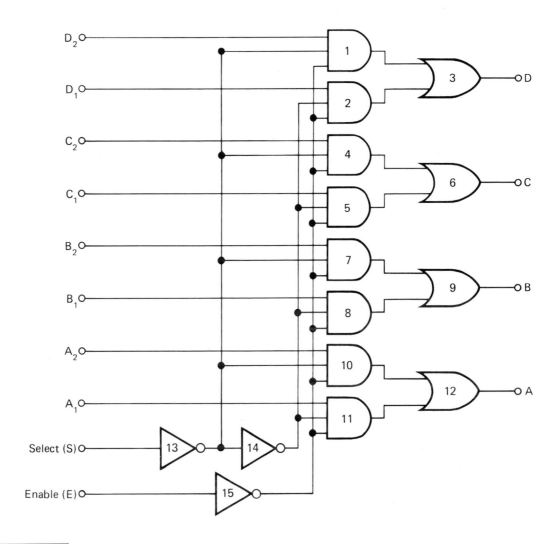

FIGURE 7–18

Quad 2-Line to 1-Line Multiplexer

Two select (address) lines allow a *4-input multiplexer*. This multiplexer selects one bit of data from four sources or four inputs. This data is selected by the select inputs S_0 and S_1. The outputs can appear in either their true or their complement (inverted) form. See Figure 7–19.

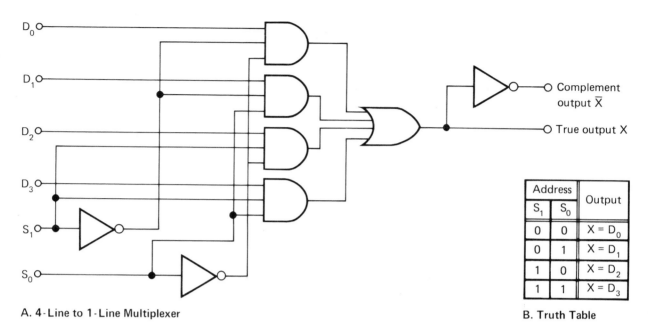

A. 4-Line to 1-Line Multiplexer

Address		Output
S_1	S_0	
0	0	$X = D_0$
0	1	$X = D_1$
1	0	$X = D_2$
1	1	$X = D_3$

B. Truth Table

FIGURE 7–19

4-Input Multiplexer

Multiplexers can be used for parallel-to-serial data conversion. For example, a 4-bit word can be serialized by using a four-to-one multiplexer and selecting one bit at a time. A *parallel-to-serial multiplexer* is shown in Figure 7–20. This circuit can be used to serialize the incoming variable data, or the input lines can be tied to ones and zeros to generate a fixed serial output word.

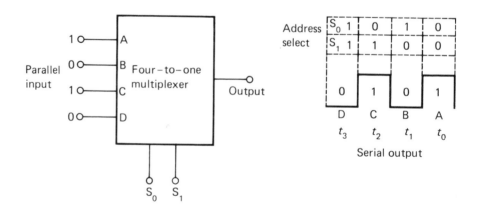

FIGURE 7–20

Parallel-to-Serial Multiplexer

DEMULTIPLEXERS

Recall that a multiplexer has multiple inputs and a single output and routes the selected or addressed data through a single output line. A *demultiplexer*, the opposite of a multiplexer, has a single input and more than one output. It routes the input data line to selected individual output lines. A demultiplexer can, therefore, be called a *data router* or *data distributor*.

A simple *2-output demultiplexer* is shown in Figure 7–21. The single input is applied to both AND gates. The select line determines whether the data appears at output 1 or output 2. Gate U_1 is enabled when the select line is high, routing the data to output 1. Gate U_2 is enabled when the select line is low, routing the data to output 2.

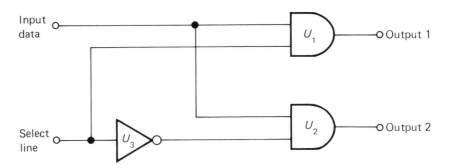

FIGURE 7–21

Simple 2-Output Demultiplexer

Figure 7–22 shows a *one-to-four demultiplexer*. The input data is output to A, B, C, or D sequentially as S_0 and S_1 change with the count. If the outputs are connected to latches and the clock occurs in synchronism with the input data, then the data distributor (demultiplexer) can work as a serial-to-parallel data converter. In this circuit, a 4-bit serial word can be converted to a 4-bit parallel word.

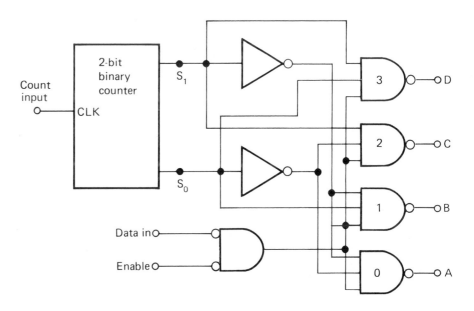

FIGURE 7–22

One-to-Four Demultiplexer

Figure 7–23 shows a multiposition switch connected as a multiplexer and as a demultiplexer. Notice that MSI multiplexers and demultiplexers work in the same way as the knob on the switch. The difference is that MSI circuits can be switched by using the address lines or select lines instead of turning the knob. These circuits are used extensively in computer systems.

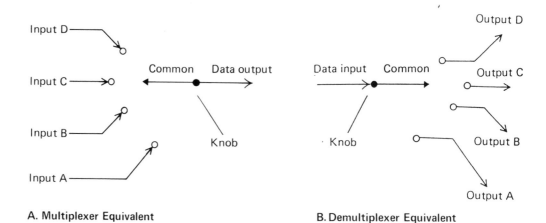

A. Multiplexer Equivalent

B. Demultiplexer Equivalent

FIGURE 7–23

**Multiposition Switch
Connected as a Multiplexer
and Demultiplexer**

**SELF-TEST
EXERCISE 7–3**

1. A _____ routes different data to the same output at different times.
2. A _____ routes the same data to different output lines depending on which line is selected.
3. A _____ converts data from serial to parallel.
4. A _____ converts data from parallel to serial.
5. A multiplexer is also called a _____.
6. A _____ has more input data lines than it has output lines.
7. A _____ has more output lines than it has input data lines.
8. Time sharing of lines is called _____.
9. Lines that use time sharing are often referred to as _____.

STOP Do Experiment 7–4

SUMMARY

An encoder converts decimal or individual inputs to a binary equivalent. It facilitates the interfacing of analog devices to digital circuits. A priority encoder is a digital device that gives priority to a particular input and ignores the other inputs until the priority input is attended to.

A decoder converts the binary code to decimal outputs that can be easily understood and used to operate analog devices. Special decoders are used to drive 7-segment displays.

A multiplexer routes any of two or more inputs to a single output. A multiplexer can be used to save space. Time sharing allows multiplexers to service many inputs. Unlike an encoder or a decoder, a multiplexer does not convert the information it receives; it simply routes the information at a desired location (input) to its output.

A demultiplexer does the same thing as a multiplexer but in reverse. It takes information at its input and outputs it to a desired location. A demultiplexer may therefore be referred to as a data distributor.

CHAPTER 7
REVIEW EXERCISES

1. Using an AND gate and as many inverters as required, draw a logic gate decoder that will recognize the number 67 in binary form.

2. Using an AND gate and as many inverters as required, draw a logic gate decoder that will detect the presence of the number 67 in BCD form.

3. How many states does hexadecimal have?

   a. 8
   b. 4
   c. 16
   d. 15
   e. 56

4. What is the maximum decimal number that can be represented by one digit in the hexadecimal number system?

5. How many input lines does a one-of-sixteen decoder have?

6. What type of combinational circuit would be used to generate a binary output code from a push-button input such as a keyboard?

7. What logic circuit is analogous to (used the same as) a single-pole mechanical selector switch?

   a. decoder
   b. multiplexer
   c. encoder
   d. exclusive OR circuit
   e. demultiplexer

8. Which of the following is not a typical application for a multiplexer?

   a. serial-to-parallel converter
   b. parallel-to-serial converter
   c. serial word generator
   d. data selector
   e. none of the above

9. The desired input data line in the experiment using the 74151 TTL multiplexer was selected by:

   a. the strobe input
   b. disabling all other inputs
   c. applying a signal to that channel
   d. a 3-bit address input code
   e. none of the above

10. Which two circuits below can be used as serial-to-parallel data converters?

   a. demultiplexer
   b. encoder
   c. multiplexer
   d. exclusive OR circuit
   e. shift register

11. If the 2 and 8 inputs were activated at the same time in a priority encoder, what would the binary output be?

12. The circuit in Figure 7–24 is a one-of-two decoder. True or False?

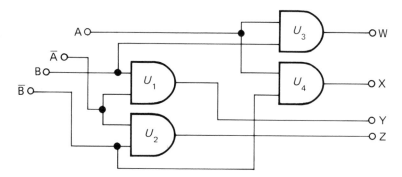

FIGURE 7–24

13. What inputs to the segments of the 7-segment common anode decoder in Figure 7–25 must be low for the decimal number 3 to be displayed?

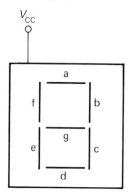

FIGURE 7–25

14. What is the data output for the address shown in Figure 7–26?

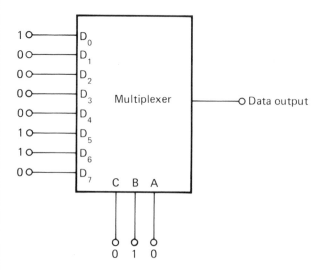

FIGURE 7–26

15. Address line is another name for select line in Figure 7–26. True or False?

EXPERIMENT 7–1 | Operation of an Encoder

PURPOSE

This experiment is designed to demonstrate the operation and characteristics of an encoder. You will build an encoder from an MSI TTL device. You will use the circuit to convert decimal inputs to binary outputs. You will also prove that the encoder is a priority encoder by simultaneously activating two or more inputs and recording the results.

EQUIPMENT

—1 digital experimenter (Equip$_1$)
—1 logic probe (Equip$_2$)
—1 74148 octal encoder IC (U_1)
—8 1 kΩ, 1/4 W resistors (R_1–R_8)
—1 8-pole dip switch (K_1)

PROCEDURE

Step 1 Connect the encoder circuit as shown in Figure 7E1–1.

NOTE: *SW$_1$* should be low to enable the chip.

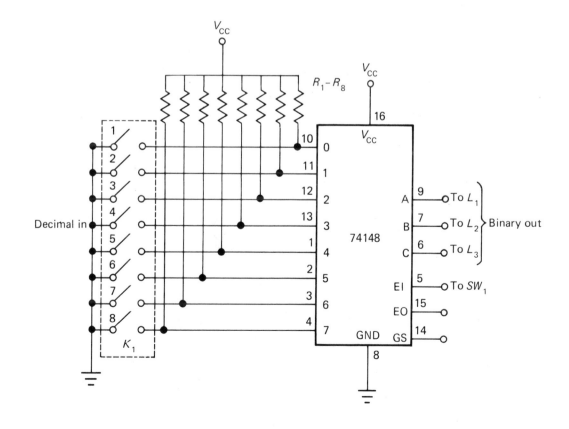

FIGURE 7E1–1

ACTIVITY _____

Apply the inputs as shown in the encoder table in Figure 7E1-2 using the 8-pole dip switch.

NOTE: The table shows the switches that should be closed; all others should be open.

EI	Input switch closed	C	B	A	GS	EO
L	1					
L	2					
L	3					
L	4					
L	5					
L	6					
L	7					
L	8					
L	2, 7					
L	All closed					
L	3, 5					
H	Try several combinations					

FIGURE 7E1-2 _____

ACTIVITY _____

Record the results as indicated by the logic indicators. Use the logic probe to check the GS and EO outputs.

NOTE: Your completed table should match the one in Figure 7E1-3. Notice that the outputs are logic lows. When more than one input is activated at the same time, the output is the binary logic low equivalent of the largest activated input.

EI	Input switch closed	C	B	A	GS	EO
L	1	H	H	H	L	H
L	2	H	H	L	L	H
L	3	H	L	H	L	H
L	4	H	L	L	L	H
L	5	L	H	H	L	H
L	6	L	H	L	L	H
L	7	L	L	H	L	H
L	8	L	L	L	L	H
L	2, 7	L	L	H	L	H
L	All closed	L	L	L	L	H
L	3, 5	L	H	H	L	H
H	Try several combinations					

FIGURE 7E1-3 _____

Step 2 Move SW_1 to the high position.

NOTE: All outputs, including GS and EO, should remain high regardless of the input condition because the inputs are not enabled.

ACTIVITY

Use the logic probe to check the GS and EO outputs.

EXPERIMENT 7–2 | Operation of a Decoder

PURPOSE

This experiment is designed to demonstrate the operation and characteristics of a decoder. You will build a simple one-of-sixteen decoder and test it for proper operation by applying all possible input conditions and recording the correctly detected output. You will also build a one-to-four decoder, test it, and record the result in a truth table. Finally, you will build an MSI TTL one-to-four decoder, test it, and record the result in a truth table.

EQUIPMENT

—1 digital experimenter ($Equip_1$)
—1 logic probe ($Equip_2$)
—1 7404 inverter IC (U_1)
—1 7421 4-input AND gate IC (U_2)
—1 74155 MSI one-to-four decoder (U_3)
—1 7408 AND gate IC (U_4)

PROCEDURE

Step 1 Construct the decoder circuit as shown in Figure 7E2–1.

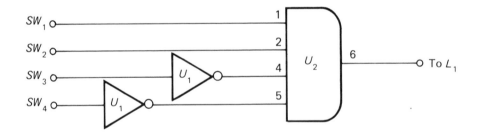

FIGURE 7E2–1

ACTIVITY

Test the circuit for all possible input conditions using the data switches. Record the results in the decoder table in Figure 7E2–2.

NOTE: The results of your tests should indicate that this circuit is a one-of-sixteen decoder.

Check: What binary input does this circuit detect?

1100_2

NOTE: Properly wired, the logic indicator will light only when the input is 1100_2. For all other states, the output should be zero or off. This circuit, therefore, detects a 12 and can be used in a counting application where a marker is needed for every twelfth item. That is, it can be used as the input to a counter that counts, not how many individual items are produced, but how many *dozen* items are produced.

Input	Output
0000	
0001	
0010	
0011	
0100	
0101	
0110	
0111	
1000	
1001	
1010	
1011	
1100	
1101	
1110	
1111	

FIGURE 7E2–2

Step 2 Carefully wire the decoder circuit as shown in Figure 7E2–3.

NOTE: Data switches SW_3 and SW_4 supply the binary inputs. The outputs are indicated by the logic indicators L_1, L_2, L_3, and L_4.

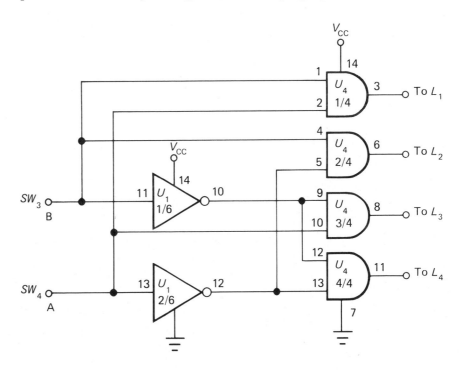

FIGURE 7E2–3

LAB

Step 3 Apply power.

ACTIVITY

Supply the inputs to the circuit as shown in the decoder table in Figure 7E2–4. Record the results as indicated by the logic indicators.

Inputs		Outputs			
B	A	L_1	L_2	L_3	L_4
SW_3	SW_4	3	2	1	0
0	0				
0	1				
1	0				
1	1				

FIGURE 7E2–4

Check: What type of circuit is this?

one-to-four decoder

Step 4 Construct the decoder circuit as shown in Figure 7E2–5.

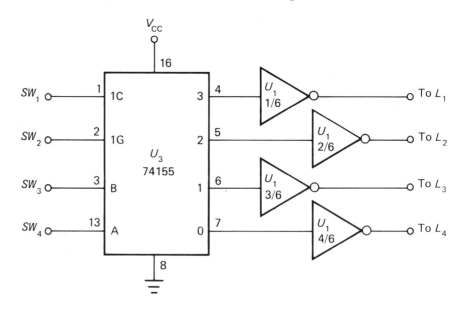

FIGURE 7E2–5

Step 5 Apply power.

NOTE: This chip contains two one-to-four decoders. The strobe input SW_2 must be low to enable the section of the chip that you are using. The data 1C input as SW_1 must be held high for one-to-four decoding. Data switches SW_3 and SW_4 supply the input data.

ACTIVITY

Apply the inputs as shown in the decoder table in Figure 7E2–6. Record the results. Then compare the results of this circuit to the results from the previous circuit.

Inputs				Outputs			
Strobe	Data	SW_3	SW_4	L_1	L_2	L_3	L_4
1G	1C	B	A	3	2	1	0
L	H	0	0				
L	H	0	1				
L	H	1	0				
L	H	1	1				

FIGURE 7E2–6 _____

Check: Do both circuits perform the same function?

Yes. They are both one-to-four decoders.

NOTE: You can see how much easier it is to wire an MSI IC as a decoder. Both circuits can be referred to as 2-line to 4-line decoders since they decode each of the four input states formed by the 2-bit inputs.

EXPERIMENT 7–3 | Operation of a 7-Segment Decoder/Driver

PURPOSE

This experiment is designed to demonstrate the operation and characteristics of a 7-segment decoder/driver. You will test a 7-segment display to determine its common connection and whether it is a common anode or a common cathode configuration.

EQUIPMENT

—1 digital experimenter (Equip$_1$)
—1 logic probe (Equip$_2$)
—1 digital multimeter with diode function (Equip$_3$)
—1 7490 BCD counter IC (U_1)
—1 7493 binary counter IC (U_2)
—1 7448 7-segment decoder/driver IC (U_3)
—1 7-segment display type FND500 or 411–819 (L_1)

PROCEDURE

NOTE: Two configurations of 7-segment displays are available: common cathode and common anode. In a *common cathode* configuration, all the segments have the negative leads tied together. To activate a common cathode, the input to a segment must be taken high. A *common anode* configuration, on the other hand, has all the positive leads tied together and must have a low input to activate the segment. The 7-segment display circuit in Figure 7E3–1 shows the internal connections for both configurations.

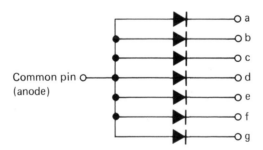

A. Common Anode

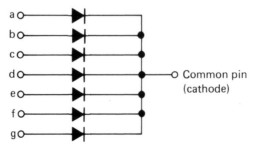

FIGURE 7E3–1

B. Common Cathode

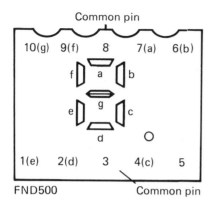

FND500

FIGURE 7E3–2

The diode function on a digital multimeter is used to test the display for this experiment. The diode function on the multimeter not only shows the junction voltage drop of a segment but also lights up a good segment. In this way, all segments can be tested before the circuit is built. To test a segment for a common cathode display, the negative lead of the multimeter is connected to the common pin, and the positive lead is connected to the input for a particular segment—for example, the d segment. If the display is indeed a common cathode and the segment is good, the d segment will light. Similarly, to test a segment for a common anode display, the positive lead of the multimeter is connected to the common pin, and the negative lead is connected to the segment input. The pin assignment diagram in Figure 7E3–2 will help you with the common cathode display connections.

Step 1 Construct the circuit as shown in Figure 7E3–3 after you determine that the display is good.

NOTE: This circuit uses a 7490 BCD counter, a 7448 decoder/driver, and a 7-segment display. The 7448 accepts the BCD input from the 7490 counter and then decodes and converts the input to a 7-segment output code that lights the proper segments for a decimal display. The 7448 has internal pull-up resistors, which eliminate the need for external resistors. Do not forget to connect the V_{CC} and ground for each chip.

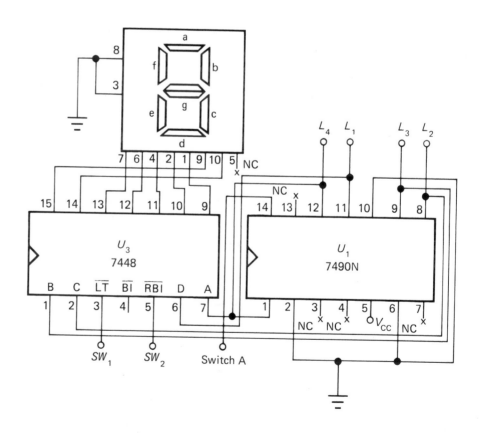

FIGURE 7E3–3

Step 2 Apply power.

Step 3 Set SW_1 low.

NOTE: A low at the lamp test input causes all segments to be on, displaying an 8 and verifying again that all segments are good.

Step 4 Set SW_1 high. The circuit is ready to count.

NOTE: Logic switch A will be used to manually apply the clock pulses.

ACTIVITY

Pulse the input with logic switch A through all possible combinations. List the displayed 7-segment outputs in the decoder table in Figure 7E3–4.

Valid input code (binary)				$\overline{RBI}$	$\overline{LT}$	Decimal number displayed
D	C	B	A			
0	0	0	0	H	H	
0	0	0	1	H	H	
0	0	1	0	H	H	
0	0	1	1	H	H	
0	1	0	0	H	H	
0	1	0	1	H	H	
0	1	1	0	H	H	
0	1	1	1	H	H	
1	0	0	0	H	H	
1	0	0	1	H	H	
1	0	1	0	H	H	
1	0	1	1	H	H	
1	1	0	0	H	H	
1	1	0	1	H	H	
1	1	1	0	H	H	
1	1	1	1	H	H	

FIGURE 7E3–4

Step 5 Set SW_4 low.

ACTIVITY

Again, pulse the input with logic switch A through all possible combinations. List the displayed outputs in the decoder table in Figure 7E3–5.

> **Check:** What difference did SW_4 ($\overline{RBI}$) make?
>
> $\overline{RBI}$ is the zero-blanking function. The zero should not have been displayed.

Step 6 Pulse the input until an 8 is displayed.

Step 7 Connect a 1 Hz clock to pin 4 of U_3, the 7448.

NOTE: The display should now pulse the displayed 8. The count can be resumed, but the display will continue to pulse.

Valid input code, (binary)				$\overline{RBI}$	$\overline{LT}$	Decimal number displayed
D	C	B	A			
0	0	0	0	L	H	
0	0	0	1	L	H	
0	0	1	0	L	H	
0	0	1	1	L	H	
0	1	0	0	L	H	
0	1	0	1	L	H	
0	1	1	0	L	H	
0	1	1	1	L	H	
1	0	0	0	L	H	
1	0	0	1	L	H	
1	0	1	0	L	H	
1	0	1	1	L	H	
1	1	0	0	L	H	
1	1	0	1	L	H	
1	1	1	0	L	H	
1	1	1	1	L	H	

FIGURE 7E3–5

Step 8 Remove the connection at pin 4. Move the clock input to the 7490 from logic switch A to a 1 Hz clock.

NOTE: This wiring will automatically increment the count every second.

Step 9 Turn off the power.

Step 10 Modify the circuit as shown in Figure 7E3–6.

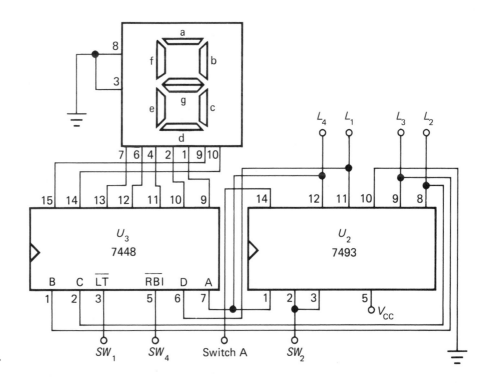

FIGURE 7E3–6

Step 11 Apply power.

Step 12 Set SW_4 and SW_1 high.

Step 13 Move SW_2 high to low to clear the counter.

ACTIVITY

Apply count pulses with logic switch A. Watch carefully after the ninth count. The 7-segment display will display the characters illustrated in Figure 7E3–7.

NOTE: Different decoder/drivers will display different characters for decimal numbers 10 through 15. Some display the hexadecimal characters A, B, C, D, E, and F; the 7448 does not.

Also, SW_4 can again be used to blank the zero by setting it low. A 1 Hz clock can be used instead of logic switch A to allow automatic counting.

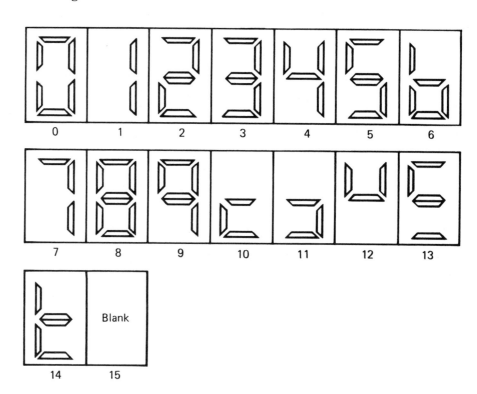

FIGURE 7E3–7

EXPERIMENT 7–4	# Operation of a Multiplexer

PURPOSE

This experiment is designed to demonstrate the operation of a multiplexer using an MSI TTL IC. You will see how a multiplexer is used to convert parallel data to serial data. You will also demonstrate how a multiplexer is used for fixed or variable serial data outputs.

EQUIPMENT

—1 digital experimenter (Equip$_1$)
—1 logic probe (Equip$_2$)
—1 74151 one-of-eight multiplexer IC (U_1)
—1 74193 binary counter IC (U_2)

PROCEDURE

Step 1 Construct the multiplexer circuit as shown in Figure 7E4–1.

NOTE: This circuit uses a 74151 one-of-eight multiplexer. Switches SW_1, SW_2, and SW_3 form the address inputs. The three logic indicators L_1, L_2, and L_3 are connected to the switches to monitor the 3-bit address code. Logic indicator L_4 is connected to the 74151 output to display the selected data. The logic probe can also be used. Wire the circuit carefully. The data input lines are connected to "0" (ground) or "1" (V_{CC}) as shown.

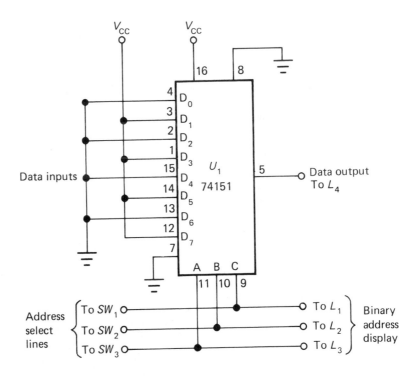

FIGURE 7E4–1

Step 2 Apply power.

ACTIVITY

Set switches SW_1, SW_2, and SW_3 as shown in the multiplexer table in Figure 7E4–2 for all possible address combinations. Record the results in the table. (Remember, L_4 is the data output indicator.)

NOTE: The binary address tells the multiplexer where to go to get the information, which, in this case, is D_0 through D_7. The information, either a constant high or a low, is then output at pin 5. Address location 000_2 is low because the data input at D_0 is low; address location 001_2 is high because the data input at D_1 is high; and so on.

Address			Data output
L_1	L_2	L_3	L_4
C	B	A	
0	0	0	
0	0	1	
0	1	0	
0	1	1	
1	0	0	
1	0	1	
1	1	0	
1	1	1	

FIGURE 7E4–2

ACTIVITY

Draw the serial output as it would be seen if the address line were clocked from 000_2 to 111_2.

NOTE: Your timing diagram should look like the one shown in Figure 7E4–3.

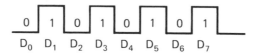

FIGURE 7E4–3

Step 3 To demonstrate how this circuit uses a 74193 counter to address the multiplexer, modify the circuit as shown in Figure 7E4–4.

NOTE: Logic switch A can be used to manually clock the counter, or the count input can be attached to a 1 Hz clock for automatic sequencing.

ACTIVITY

Fill in the table in Figure 7E4–5 as you did for the previous multiplexer circuit.

NOTE: Now you can use switches SW_1, SW_2, SW_3, and SW_4 for the data lines instead of for the address lines since the 74193 is addressing the multiplexer.

ACTIVITY

Draw the timing diagram for this output.

NOTE: Your timing diagram should still look like the one shown in Figure 7E4–3. All you have done is added a digital counter to the circuit so that the switches are not needed to input the binary address.

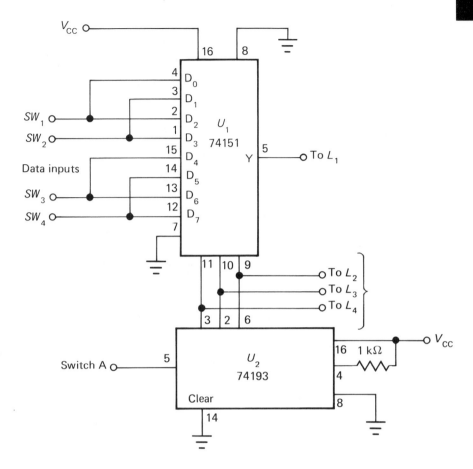

FIGURE 7E4—4

FIGURE 7E4—5

Address			Data output
L_1	L_2	L_3	L_4
C	B	A	
0	0	0	
0	0	1	
0	1	0	
0	1	1	
1	0	0	
1	0	1	
1	1	0	
1	1	1	

Step 4 To demonstrate how this circuit can be used as a parallel-to-serial data converter (converting an 8-bit parallel data word to an 8-bit serial data word), change any input data D_0 to D_7 and watch how the output changes accordingly. In this way, the inputs can be used to set a fixed serial output.

ACTIVITY

Figure 7E4–6 shows a fixed output signal. Using what you already know about multiplexers, create the output shown in this figure. *Hint:* You can view the output on an oscilloscope if you use a 100 kHz clock signal for the counter.

NOTE: If you have successfully created the required fixed output, D_0 through D_7 should be as shown in the table in Figure 7E4–7.

FIGURE 7E4–6

FIGURE 7E4–7

D_0	1
D_1	1
D_2	0
D_3	1
D_4	1
D_5	0
D_6	1
D_7	1

Exclusive OR and NOR Circuits

OBJECTIVES

After studying this chapter, you will be able to:

1. Draw the truth table for an Exclusive OR gate.
2. Draw the logic symbol and logic equation for an Exclusive OR gate.
3. Describe the operation of Exclusive OR and Exclusive NOR gates.
4. Identify the logic function referred to as a comparator.
5. Explain the operation, use, and limitations of a parity generator and a parity checker.
6. Identify a binary word as having odd or even parity.
7. Add simple binary numbers.
8. Describe the difference between a full adder and a half adder.
9. List the steps in binary addition.
10. Identify the number of full or half adders needed to add two given binary numbers.

INTRODUCTION

Exclusive OR and NOR circuits may be considered the most versatile of the digital circuits and therefore deserve special recognition. This chapter focuses on two particular applications of these circuits. The parity circuits for which Exclusive OR and NOR gates are used are the sentries of digital communications. These gates are also used to make adder circuits.

EXCLUSIVE OR GATE CIRCUIT

The Exclusive OR gate has many applications in digital systems. It is a 2-input logic circuit that produces a binary one output when only one input is binary one. Figure 8–1 shows the truth table, the logic symbol, and the equivalent circuit for the Exclusive OR gate. Notice in Figure 8–1 that the light comes on only when one switch is up (high) and the other is down (low).

The logic equation for the Exclusive OR gate, as derived from the truth table, is $C = \overline{A}B + A\overline{B}$. A special symbol, $\oplus$, is used for this function. Therefore, $C = A \oplus B$ indicates the Exclusive OR function.

Exclusive OR circuits are available in MSI form and do not need to be constructed from basic gates. The 7486 is an example of a MSI TTL

Exclusive OR circuit. Its truth table and pin assignments are shown in Figure 8–2.

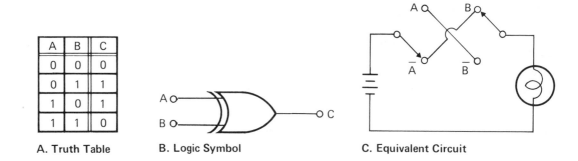

A	B	C
0	0	0
0	1	1
1	0	1
1	1	0

A. Truth Table

B. Logic Symbol

C. Equivalent Circuit

FIGURE 8–1

Exclusive OR Gate

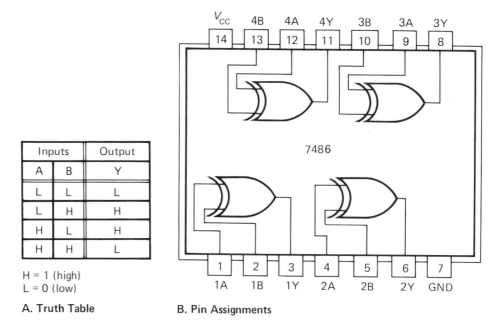

Inputs		Output
A	B	Y
L	L	L
L	H	H
H	L	H
H	H	L

H = 1 (high)
L = 0 (low)

A. Truth Table

B. Pin Assignments

FIGURE 8–2

7486 Exclusive OR Circuit

EXCLUSIVE NOR GATE CIRCUIT

The Exclusive NOR gate is the inverse of the Exclusive OR gate. Figure 8–3 shows the truth table, the logic symbol, and the logic equation for this function. The Exclusive NOR gate is sometimes referred to as a *comparator*. (Comparators will be discussed in more detail later in this chapter.)

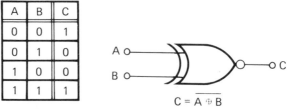

A	B	C
0	0	1
0	1	0
1	0	0
1	1	1

FIGURE 8–3

Exclusive NOR Gate

A. Truth Table

B. Logic Symbol and Equation

$C = \overline{A \oplus B}$

SELF-TEST EXERCISE 8–1

1. The Exclusive OR gate output is high:
   a. when both inputs are binary one
   b. when only one input is binary one
   c. when both inputs are binary zero
   d. all the above
   e. none of the above
2. Draw the logic symbol for the Exclusive OR gate.
3. Draw the truth table for the Exclusive OR gate.
4. The _____ is the inverse of the Exclusive OR gate.
5. The _____ is sometimes referred to as a comparator.
6. Draw the truth table for the Exclusive NOR gate.
7. Draw the logic symbol for the Exclusive NOR gate.

 Do Experiment 8–1

EXCLUSIVE OR APPLICATIONS

As complex and fast as digital circuits are, they are susceptible to errors. Errors due to noise or other factors are possible. When digital information is transmitted within a system from location to location, pulses may be accidentally lost or added. If these errors can be detected the moment they occur, they are less likely to cause a significant problem. The use of a *parity bit* helps in the detection of this loss or addition of a bit in data transmission from one place to another.

Parity is designated as either even or odd parity. In a parity generator circuit, a parity bit is added either at the beginning or at the end of the transmitted data word, depending on design. A parity bit is transmitted with the data either in parallel or series, depending on the system. When the binary data word and parity bit are received, they are tested for bit errors in a parity checker circuit. These common applications of the Exclusive OR gate—the parity generator and the parity checker—are described next, and examples are given to demonstrate their operation.

Parity Generator

A *parity generator* generates an output that indicates an even or odd number of bits in a word. The devices receiving and transmitting data must agree on the selection of odd or even parity prior to data being sent.

In *odd parity*, the total number of high bits in a word must be odd. If the binary data word to be transmitted contains an even number of ones, then a one is placed at the parity position. This parity bit makes the total number of ones in the transmitted word odd. If the binary data word to be transmitted contains an odd number of high bits, then a zero is placed at the parity position, which leaves the number of ones in the word odd. The parity position can be either the LSB or the MSB of the transmitted word. This new word is then transmitted. For example, in the ASCII code, the letter K is represented by the 7-bit binary number 100 1011. If an odd parity bit were added to the MSB location of this data word, the transmitted word would be 1100 1011. In *even parity*, a parity bit is generated to make the total number of ones even. With an even parity bit, the ASCII character K would be transmitted as 0100 1011. More examples are listed in Table 8–1.

TABLE 8–1

Examples of ASCII Codes with Parity Bits

Character	ASCII Code	Odd Parity	Even Parity
A	100 0001	1100 0001	0100 0001
6	011 0110	1011 0110	0011 0110
<	011 1100	1011 1100	0011 1100
N	100 1110	1100 1110	0100 1110
Z	101 1010	1101 1010	0101 1010
%	010 0101	0010 0101	1010 0101
1	011 0001	0011 0001	1011 0001

Parity Checker

While a parity bit is generated at the transmitting (sending) end of a digital system, examination for loss or addition of a bit occurs at the receiving end of the system. That is, the binary word is examined when it reaches the receiving end, and another parity bit is generated for the transmitted data word. This newly generated parity bit is checked against the transmitted parity bit. If the bits are not the same, an error is indicated. A *parity checker* is an Exclusive OR gate circuit that is used to determine whether or not two parity bits are the same.

■ **EXAMPLE**

A 4-bit parallel word parity generator is shown in Figure 8–4. Notice that as the 4-bit data word 0101 is being transmitted, the output for even parity is a zero, and the transmitted word would therefore be 00101. For odd parity, the transmitted word would be 10101.

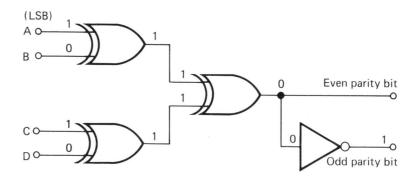

FIGURE 8–4

4-Bit Parallel Word Parity Generator

■ **EXAMPLE**

Figure 8–5 is another example of a 4-bit parallel word parity generator.

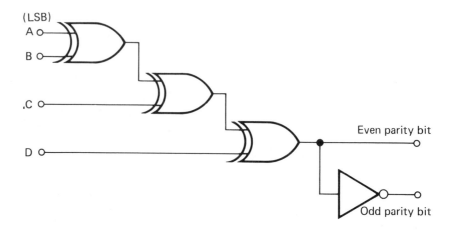

FIGURE 8–5

Another 4-Bit Parallel Word Parity Generator

Check the operation of this circuit using the inputs shown in Figure 8–4. You can see that the operation is basically the same. ∎

■ **EXAMPLE**

Figure 8–6 shows a 4-bit serial word parity generator. Notice in this example that as the data word is being transmitted, a pulse is sent to R to clear the register (U_3) for the start of the word (Q goes low). Each high bit pulse on the word line toggles the register; therefore, if the number of toggle pulses (one bits) is odd, the register Q line will be high at the end of the word. If the number of one bits is even, the Q line will be low. This bit, high or low, is one of the inputs to the AND gate (U_2). The end of the word pulse will enable this AND gate, which will then pass the parity bit (the Q output) to the OR gate (U_1). U_1 will send this parity bit out as part of the transmitted word.

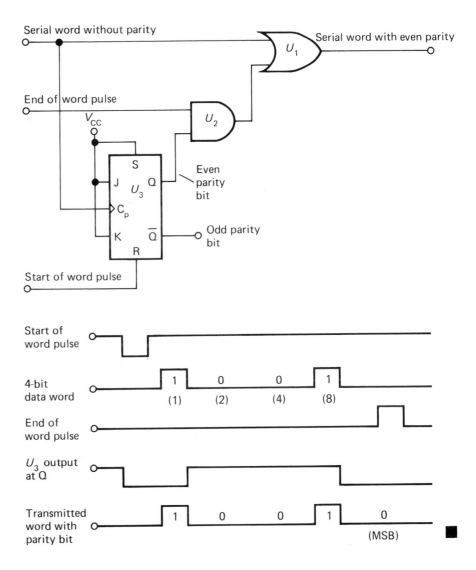

FIGURE 8–6

4-Bit Serial Word Parity Generator

■ **EXAMPLE**

Figure 8–7 shows a 4-bit parity checker circuit. In this circuit, the incoming data word is sent through a parity generator (U_1, U_2, and U_3) that generates a parity bit for the data word it received. Then, at U_4, this newly generated parity bit is compared with the one received with the transmitted word. If the parity bits are the same, the data is assumed to be correct and no error message is generated. If they are not the same, U_4 will output a high, indicating an error. Although this circuit cannot detect all multiple-bit errors, it is inexpensive, reliable, and commonly used.

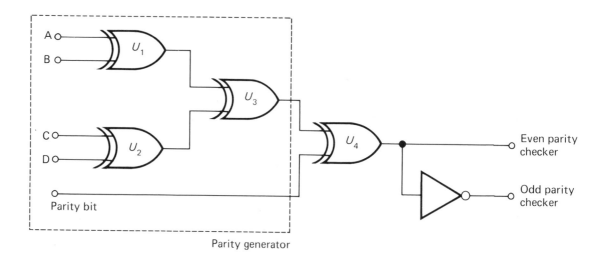

Parity generator

FIGURE 8–7

4-Bit Parity Checker

SELF-TEST EXERCISE 8–2

1. The parity generator and parity checker are common applications for the _____.

2. A parity bit can be transmitted as the _____ or _____ bit in a word.

3. In even parity, the total number of ones in the transmitted word is _____.

4. In odd parity, the total number of ones in the transmitted word is _____.

5. Determine the even parity bit for the following words:
   a. __ 111 1111
   b. __ 101 1001
   c. __ 011 0000
   d. __ 000 0000
   e. __ 100 1100

6. Determine the odd parity bit for the following words:
   a. __ 110 1011
   b. __ 101 0011
   c. __ 000 0000
   d. __ 101 0101
   e. __ 111 0101

7. A parity checker outputs a high when _____ in transmission is detected.

8. A parity checker cannot detect _____ in transmission.

STOP   Do Experiment 8–2

EXCLUSIVE NOR APPLICATION

Comparators

Exclusive NOR circuits can be used to build *comparators*. Each Exclusive NOR gate is a single-bit comparator. If both bits to be compared are equal, it gives a high output; otherwise, it outputs a low. Four or eight Exclusive NOR gates can be used to compare two 4- or 8-bit binary words to determine whether or not they are equal. The 4- or 8-bit outputs from the Exclusive NOR gates are ANDed together to produce a single output.

A typical 4-bit binary word comparator is shown in Figure 8–8. This circuit can be modified with additional components to give an output if word A is less than B or if word A is greater than B. The 74LS266 TTL MSI Exclusive NOR circuit shown in Figure 8–9 can be used to build a 4-bit digital comparator.

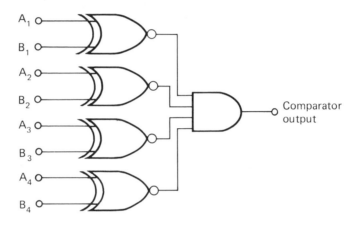

FIGURE 8–8

4-Bit Binary Word Comparator

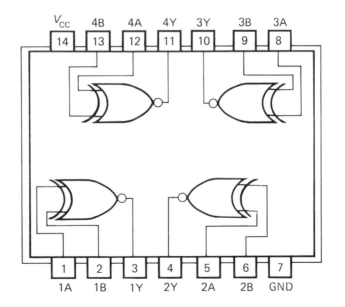

FIGURE 8–9

74LS266 Pin Assignments

BINARY ADDERS

A *binary adder* is a circuit that adds two binary numbers. The adder is the main functional element of digital computers. The adder circuit is also used in calculators and other digital instruments that perform mathematical functions.

Binary Addition

The basic steps for binary addition are very similar to the steps for decimal addition. Look at the example in Figure 8–10A. The decimal numbers 52 and 71 are added by beginning with the LSB. The 2 and 1 are added; since there is no carry, the 3 is recorded. Next, the tens column is added: $7 + 5 = 12$. The 2 is recorded in the tens column and the 1 carried to the next column. The sum is 123.

Binary addition is done in the same way. In a binary adder, the numbers are always added by starting at the LSB location. Look at the example in Figure 8–10B. The numbers 12 and 15, which in binary are 1100 and 1111, are added as follows:

1. The two least significant bits are added first: $0 + 1 = 1$, no carry.
2. The next two bits are added: $0 + 1 = 1$, no carry.
3. The next two bits are added: $1 + 1 = 2$, which in binary is 10. When binary sums have more than one bit, a carry to the next bit is generated. The bit must be added to the next two bits.
4. This step requires the addition of three bits because of the carry: $1 + 1 + 1 = 3$, which in binary is 11. The 1 is recorded and the other 1 carried. The sum is a 5-bit number; the carry bit represents the MSB.

More examples of binary addition are shown in Figure 8–11. Notice that fractions are added by first lining up the binary point (same as the decimal point in the decimal system) and then adding as usual.

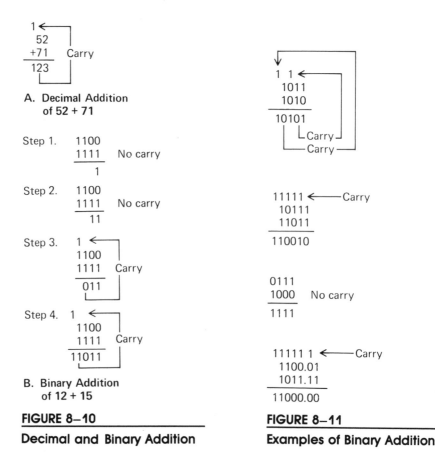

A. Decimal Addition
of 52 + 71

B. Binary Addition
of 12 + 15

FIGURE 8–10

Decimal and Binary Addition

FIGURE 8–11

Examples of Binary Addition

Further examination of the results of binary addition reveals that the result of adding any two binary digits is the same as the truth table values for an Exclusive OR circuit, except for the carry bit. See Figure 8–12.

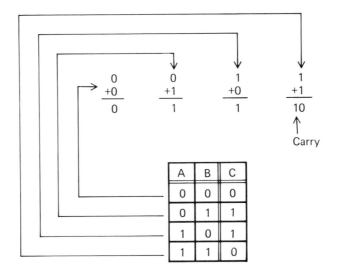

FIGURE 8–12

Exclusive OR Truth Table Related to Results of Binary Addition

Half Adders and Full Adders

Circuits that perform addition functions can be constructed with Exclusive OR and AND gates. If circuitry is added to create a carry out, a single-bit adder is constructed. See Figure 8–13. When the LSB of the two numbers is added, a carry is generated that must be added to the next two bits. Another circuit must be built that can accept this carry bit and add it to the next two numbers. The circuit shown in Figure 8–13 does not have the ability to accept a carry and, therefore, is a half adder. A *half adder* is a logic circuit that adds two bits and produces two binary digits on its output, one sum bit and one carry bit.

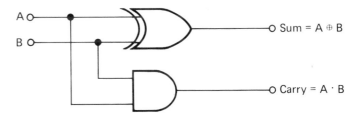

FIGURE 8–13

Single-Bit Half Adder Circuit

The circuit shown in Figure 8–14 is a full adder. A *full adder* is a logic circuit that adds three bits and produces two bits on its output, one sum bit and one carry bit. In Figure 8–14, gates U_1 and U_2 add bits A_1 and B_1, producing a sum (S_1) and a carry (C_1). Gate U_3 adds the carry from the previous stage (if there is one) to the sum S_1 and produces the final sum. Gate U_4 gives the carry out for the sum of A_1 and B_1 and the carry from the previous stage (if there is one). These two carry outputs are ORed together at U_5 to produce the final carry output, which will feed the next significant bit adder in a multibit adder.

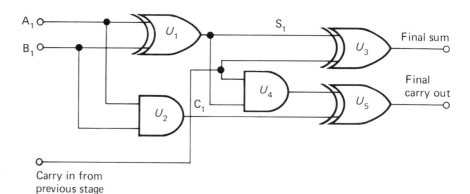

FIGURE 8–14

Single-Bit Full Adder Circuit

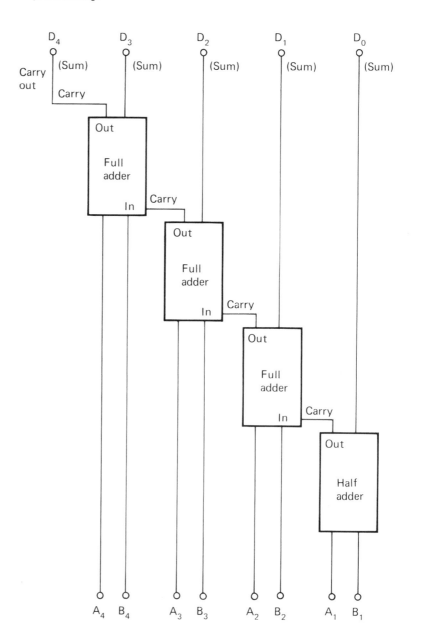

FIGURE 8–15

Block Diagram of Adder Circuit for Two 4-Bit Binary Numbers

Full adders and half adders are cascaded to form adders for two multibit numbers. Figure 8–15 is a block diagram of an adder circuit for two 4-bit binary numbers (A and B). As shown, A consists of the 4-bit number A_3, A_2, A_1, and A_0; B consists of the 4-bit number B_3, B_2, B_1, and B_0. Each of the corresponding bits of the two numbers is added in the circuit from the LSB to the MSB. The LSBs, A_0 and B_0, are added in a half adder (because there is no carry in possible). This sum is output at D_0. The carry is cascaded to the next stage, which adds this carry bit to the next two bits, A_1 and B_1, until the sums and final carry are completed. The result is the 5-bit output D_4, D_3, D_2, D_1, and D_0. Note, that since adders too are available in MSI form, there is no need to construct them from individual gates.

SELF-TEST EXERCISE 8–3

1. The _____ circuit is used to compare words to determine whether or not they are equal.
2. Add the following binary words:

   a. 1101
       1010

   b. 1111
       0011

   c. 1011011
       0101011

   d. 1111
       0000

   e. 101010
       010101

   f. 00.01
       10.00

3. A half adder has a (carry in, carry out).
4. A half adder does not have a (carry in, carry out).
5. How many full adders and how many half adders are needed to add an 8-bit number?
6. In a parallel binary adder:
   a. all bits are added simultaneously
   b. all bits are added sequentially, LSB first
   c. all bits are added sequentially, MSB first
   d. the bits cannot be added
   e. none of the above

 Do Experiment 8–3

SUMMARY

The Exclusive OR gate is a 2-input logic circuit that produces a binary one at its output only when the inputs are not the same—that is, when one input is a zero and one input is a one. The Exclusive NOR gate is similar, except that it outputs a high when the inputs are the same—that is, when both inputs are ones or both are zeros. It is also called a comparator.

    Parity is used to detect the addition or deletion of a bit in data transmission. Although it is not foolproof, it is an economical method for detecting most errors. In even parity, the total number of ones in a word is even. In odd parity, the total number of ones in a word is odd. Usually, one method, even or odd, is chosen for the detector. Often, however, a computer is capable of producing even or odd transmission. In this case, the receiving detector will inquire of the sender prior to transmission whether even or odd parity is to be expected.

A binary adder, constructed from Exclusive OR or NOR circuits, adds binary numbers. The circuits are usually part of an IC made for the specific function of addition. Regardless of what type of circuit is used, the numbers must be added by starting with the LSB. A full adder is a logic circuit that has both a carry in and a carry out. A half adder is a logic circuit that does not have a carry in, although it does have a carry out.

CHAPTER 8 REVIEW EXERCISES

1. Draw the logic symbol and truth table for the Exclusive OR gate.

2. Mark the following binary words as having odd or even parity:
   a. 001101 _____
   b. 01100110 _____
   c. 01010101 _____
   d. 10101 _____
   e. 111 _____
   f. 0001 _____
   g. 01010 _____
   h. 00 _____

3. The number of ones in a transmitted word should be _____ in even parity.
   a. odd
   b. even
   c. 5
   d. 1
   e. doesn't matter

4. Another name for Exclusive NOR is:
   a. NOR
   b. sum of products
   c. comparator
   d. NOT-OR
   e. parity

5. Add the following numbers using binary addition:
   a. 1011100
      0101101
   b. 00110
      01111
   c. 10110
      10110
   d. 111.11
      011.01

e. 000011111
   111100000

f. 0101010101
   0111001101
      110101

6. In a parallel binary adder:

   a. all bits are added simultaneously
   b. all bits are added sequentially, LSB first
   c. all bits are added sequentially, MSB first
   d. the bits cannot be added
   e. none of the above

7. A half adder is:

   a. a single-bit adder with carry in
   b. a single-bit adder with carry out but no carry in
   c. one-half of a full adder
   d. a single-bit adder with no carry in or out
   e. the same as a full adder

8. How many full adders and half adders are needed to add two 8-bit numbers?

   a. 6 full and 2 half
   b. 7 half and 1 full
   c. 7 full and 1 half
   d. 4 of each
   e. 8 half adders

9. Draw the block diagram of an adder circuit that can add 01011 + 0011 = 1110 (11 + 3 = 14). Label the blocks as full or half adders and show the sum and carry lines.

| EXPERIMENT 8–1 | # Operation of the Exclusive OR Gate |

PURPOSE

This experiment is designed to show the operation and characteristics of the Exclusive OR gate. You will build an Exclusive OR gate circuit using a 7486 IC. You will then test the circuit and complete the truth table for this operation.

EQUIPMENT

—1 digital experimenter (Equip$_1$)
—1 logic probe (Equip$_2$)
—1 7486 Exclusive OR gate (U_1)

PROCEDURE

Step 1 Construct the Exclusive OR circuit as shown in Figure 8E1–1.

NOTE: Data switches SW_1 and SW_2 are used as the inputs. Logic indicator L_1 is used to represent the output. There are four Exclusive OR gates in a 7486 IC. This is only one of them. Be sure to apply power to the IC. Either use a TTL data book for the pin assignments or use Figure 8–2.

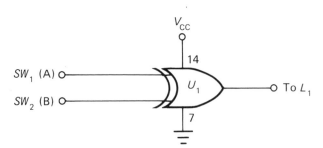

FIGURE 8E1–1

Step 2 Apply power.

ACTIVITY

Test the circuit for all possible input combinations and complete the truth table in Figure 8E1–2.

A	B	L_1
0	0	
0	1	
1	0	
1	1	

FIGURE 8E1–2

EXPERIMENT 8–2

Operation of a Parity Generator and Checker

PURPOSE

This experiment is designed to demonstrate the operation of a parity generator and a parity checker. You will construct each circuit and also determine the odd and even parity bit for given binary words.

EQUIPMENT

—1 digital experimenter (Equip$_1$)
—1 logic probe (Equip$_2$)
—1 7486 Exclusive OR IC (U_1)
—1 7404 inverter IC (U_2)

PROCEDURE

Step 1 Construct the parity generator as shown in Figure 8E2–1.

NOTE: Either use a TTL data book for the pin assignments or use the pin assignments at the beginning of this chapter. Be sure to apply power to the IC. Use the data switches SW_1, SW_2, SW_3, and SW_4 for the inputs.

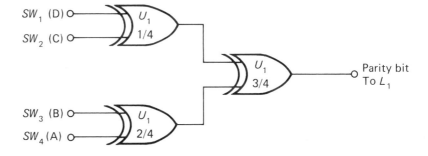

FIGURE 8E2–1

Step 2 Apply power.

ACTIVITY

Set the switches to all possible input conditions and complete the truth table in Figure 8E2–2. Logic indicator L_1 can be used to represent the parity bit if desired.

Check: Is this even or odd parity?

This is even parity because the total number of ones in the word is even. The output of the parity generator can be inverted to produce odd parity by using U_2 to invert the output of the circuit.

Binary data word				Parity bit
SW_1 (D)	SW_2 (C)	SW_3 (B)	SW_4 (A)	
0	0	0	0	
0	0	0	1	
0	0	1	0	
0	0	1	1	
0	1	0	0	
0	1	0	1	
0	1	1	0	
0	1	1	1	
1	0	0	0	
1	0	0	1	
1	0	1	0	
1	0	1	1	
1	1	0	0	
1	1	0	1	
1	1	1	0	
1	1	1	1	

FIGURE 8E2–2

NOTE: The truth table in Figure 8E2–3 shows both the even and the odd parity bit for the given words.

Binary data word				Even parity bit	Odd parity bit
SW_1 (D)	SW_2 (C)	SW_3 (B)	SW_4 (A)		
0	0	0	0	0	1
0	0	0	1	1	0
0	0	1	0	1	0
0	0	1	1	0	1
0	1	0	0	1	0
0	1	0	1	0	1
0	1	1	0	0	1
0	1	1	1	1	0
1	0	0	0	1	0
1	0	0	1	0	1
1	0	1	0	0	1
1	0	1	1	1	0
1	1	0	0	0	1
1	1	0	1	1	0
1	1	1	0	1	0
1	1	1	1	0	1

FIGURE 8E2–3

Step 3 Modify the circuit as shown in Figure 8E2–4.

NOTE: This circuit is a 4-bit parity checker. The transmitted 4-bit data word (data switches SW_1, SW_2, SW_3, and SW_4) generates a parity bit at logic indicator L_2. This bit is compared to the transmitted (fifth line) parity bit (logic switch $\overline{A}$) at logic indicator L_3. L_2 and L_3 will be the same if the word transmitted was correctly received. If they are not, an error has occurred and L_1 will light.

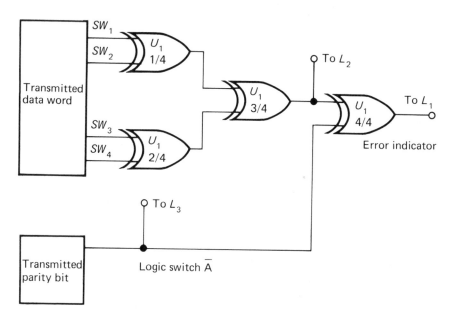

FIGURE 8E2–4

ACTIVITY

Test each input from the table in Figure 8E2–2 using logic switch $\overline{A}$ to send the transmitted parity bit.

NOTE: L_1 will light when the parity bit sent by logic switch $\overline{A}$ is not the correct parity bit for the data word at the switches. Remember, the switches represent the transmitted data word (as received). L_2 and L_3 must be the same, or L_1 will light to indicate an error.

EXPERIMENT 8–3

Operation of an Adder Circuit

PURPOSE

This experiment is designed to demonstrate the operation of a simple adder circuit. You will build this circuit using a 7482 IC. You will then test the operation of the circuit by inputting 2-bit binary words and recording the output.

EQUIPMENT

—1 digital experimenter (Equip$_1$)
—1 logic probe (Equip$_2$)
—1 7482 2-bit adder (U_1)

PROCEDURE

Step 1 Construct the circuit as shown in Figure 8E3–1.

NOTE: This circuit is a full adder because it has a carry in (pin 5). But, it is being used as a half adder in this circuit since the carry in is not used. Data switches SW_1, SW_2, SW_3, and SW_4 are used to input the binary numbers to be added. Logic indicators L_1, L_2, and L_3 are used to represent the sum of the numbers.

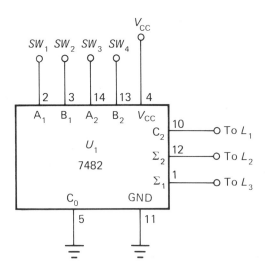

FIGURE 8E3–1

ACTIVITY

Add the binary words in the truth table in Figure 8E3–2 and record the results.

ACTIVITY

Figure 8E3–3 shows the addition problems from the truth table written as you would ordinarily see them. Compare your results from the table to the answers in Figure 8E3–3.

Inputs				Outputs		
A_1	B_1	A_2	B_2	Σ_1	Σ_2	C_2
L	L	L	L			
H	L	L	L			
L	H	L	L			
H	H	L	L			
L	L	H	L			
H	L	H	L			
L	H	H	L			
H	H	H	L			
L	L	L	H			
H	L	L	H			
L	H	L	H			
H	H	L	H			
L	L	H	H			
H	L	H	H			
L	H	H	H			
H	H	H	H			

FIGURE 8E3–2

$$A^2 A^1$$
$$B^2 B^1$$
$$\overline{\Sigma^2 \Sigma^1}$$

00	01	00	01	10	11
00	00	01	01	00	00
000	001	001	010	010	011

10	11	00	01	00	01
01	01	10	10	11	11
011	100	010	011	011	100

10	11	10	11
10	10	11	11
100	101	101	110

FIGURE 8E3–3

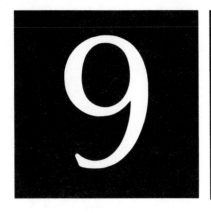

9 Digital-to-Analog and Analog-to-Digital Circuits

OBJECTIVES

After studying this chapter, you will be able to:

1. Draw the schematic symbol for an operational amplifier.
2. Define an op amp, a DAC, and an ADC.
3. Define input resistance, output resistance, gain, and feedback.
4. Calculate the gain of a negative or a positive feedback op amp circuit given R_f and R.
5. Calculate the output voltage for a summing amplifier given V_{in}, R_f, and R.
6. Describe the comparator operation of an op amp.
7. Determine V_{out} for a DAC circuit given the digital input and LSB resolution.
8. Determine V_{out} for a DAC circuit given the binary input word, V_{ref}, R_f, and R.
9. Calculate the binary input required for a specified V_{out} given the DAC bit resolution and V_{ref}.
10. Define a transducer.
11. Define start conversion and end conversion.
12. Calculate the digital output for an analog input given the LSB resolution of the ADC.
13. Calculate the conversion time for an N-bit word given the ADC bit resolution and the clock frequency for a digital ramp and a successive-approximation ADC.

INTRODUCTION

Analog-to-digital and digital-to-analog circuits allow digital circuits to work for us. The digital-to-analog circuits make it possible to convert very small digital signals to some other form that can be used to control, for example, industrial equipment or home appliances. The analog-to-digital circuits convert signals from this equipment to a form that digital circuits can manage.

Digital circuits use only ones and zeros (on or off inputs). Therefore, to use digital circuits for many applications, conversion of analog signals to digital signals is necessary for processing. Once the signals are in binary form, a controller or computer can process the information. A digital (binary) signal must often be converted back to an analog signal after processing to perform a useful function, such as making an adjustment in a system that requires corrective action.

An *analog-to-digital converter* (ADC) is used to convert analog signals to digital equivalent values. A *digital-to-analog converter* (DAC) is used to convert digital signals to analog equivalent values. Although these converters can be purchased in monolithic IC form (monolithic means that the individual components—the resistors, transistors, diodes, and so on—are placed on a single layer of silicon material in the IC), in this chapter we will analyze the circuits in discrete form to better understand how they work.

OPERATIONAL AMPLIFIERS

The most important part of a digital converter is the *operational amplifier* (op amp). An amplifier controls a large amount of voltage, current, or power by a smaller amount of voltage, current, or power. An amplifier does not originate any power. It is a control device; it controls the amount of available power that is routed to the circuit. An op amp is a special type of high-gain direct current (dc) amplifier.

The op amp was derived from a group of high-performance dc amplifiers originally designed for use in analog computers to perform mathematical operations. It consists of several amplifier stages cascaded together. Each of these stages has specific characteristics that distinguish the op amp from an ordinary amplifier. Op amps are usually constructed in IC form. Each op amp is a complete amplifier system having very high amplification (gain), high input resistance, and very low output resistance. An op amp may be used to amplify and/or manipulate the signals processed in a digital-to-analog or analog-to-digital converter.

Standard Symbol

Figure 9–1 shows the standard schematic symbol for an op amp IC. Notice how this symbol differs from an inverter symbol. The op amp symbol remains the same regardless of the type of op amp, package material, or package configuration. Supply voltages are shown here but are often omitted from schematics. An example of a typical, widely available, and low cost op amp is the 741 op amp.

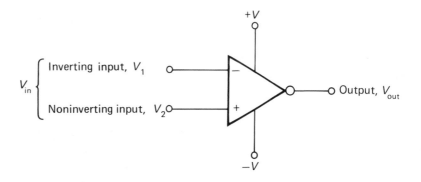

FIGURE 9–1

Standard Schematic Symbol of an Op Amp

Internal Stages

Figure 9–2 shows the internal stages of a typical op amp. The first stage in almost all op amps is a *differential amplifier*. A differential amplifier has two inputs and one output. The output is the amplified difference between the two input signals. The output is in phase with the positive input and 180° out of phase with the negative input.

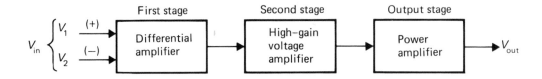

FIGURE 9–2

Internal Stages of an Op Amp

Differential amplifiers have some gain, but to provide more gain, a second stage, or *high-gain voltage amplifier*, is added, which boosts the overall gain to its high value. This stage may be composed of several transistors, which are often connected in Darlington pairs. A typical op amp may have a voltage gain of 200,000 or more.

The third, or output, stage is almost invariably a *power amplifier*. Its primary function is to supply power to the load at a low impedance. The output stage allows the op amp to deliver several milliamperes (mA) of current to the load.

Pin Assignment

The pin assignment for the 741 8-pin DIP (dual in-line package) is shown in Figure 9–3. Pins 4 and 7 are the power supply inputs: Pin 7 is the $+V_{CC}$; pin 4 is the $-V_{CC}$. The op amp is often powered by both negative and positive power supplies, which allows the output to swing negative or positive with respect to ground. Most op amps require supply voltages for $+V_{CC}$ of 5 V to 18 to 20 V and for $-V_{CC}$ of ground to -18 to -20 V.

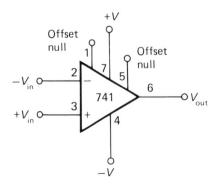

FIGURE 9–3

Pin Assignment for 741 8-Pin DIP

Pins 2 and 3 are the input signal terminals: Pin 2 is the negative input; pin 3 is the positive input. The negative input is the inverting input; the output is 180° out of phase with this input. The positive input is used if the output is in phase with the input. If two different voltages are applied to these inputs, the output is the amplified difference between the two.

Pin 6 is the output. Pins 1 and 5 are the offset null inputs. These inputs allow an external variable resistor to be connected to the circuit. The resistor is set so that the output of the op amp is exactly 0 V when the two input terminals are at 0 V, which compensates for any undesired difference in the circuit.

Electrical Characteristics

Input Resistance. *Input resistance* (R_{in}) is the resistance seen by a signal "looking into" the amplifier input terminals. Generally, the higher the input resistance, the better the op amp will perform. An input resistance of 2 MΩ is typical.

Output Resistance. *Output resistance* (R_{out}) is the resistance seen by the load "looking back into" the output terminals. The lower the output resistance, the better. Less than 100 Ω is typical.

Input Capacitance. *Input capacitance* (C_{in}) is the capacitance at one of the input terminals when the other terminal is grounded. The input capacitance can affect circuit operation at high frequencies. Less than 2 pF (picofarads) is typical.

Gain. *Gain* (A_v) is the voltage gain of the amplifier in the open-loop (V_{ol}) mode of operation. Open-loop gain (A_{vo}) is gain with no feedback. A gain of 200,000 or higher is typical. Open-loop gain is seldom used because of stability problems. An op amp is usually operated with negative feedback. The gain characteristic is often frequency sensitive, which results in decreased gain at various frequencies.

Common-Mode Rejection Ratio. The *common-mode rejection ratio* (CMRR) is the ratio of the differential voltage gain to the common-mode voltage gain. That is, an op amp has the ability to amplify only the difference between two signals applied to the inputs and not the signal that is the same on both inputs. Unwanted signals like noise and hum are thereby rejected.

Slew Rate. The *slew rate* indicates how fast the output voltage can change. Slew rate is expressed in volts per second (V/s). A slew rate of 1 V/μs means that the output voltage can change no faster than one volt each microsecond, and thus the input voltage can change no faster than that for accurate operation. The slew rate determines how long a signal must be applied to the input in a steady state before an output is read.

SELF-TEST EXERCISE 9–1

1. An ADC is:
    a. a digital-to-analog converter
    b. an analog-to-digital converter
    c. none of the above

2. A DAC is:
   a. a digital-to-analog converter
   b. an analog-to-digital converter
   c. none of the above
3. An op amp is a special type of (high-, low-) gain dc amplifier.
4. Draw the standard symbol for an op amp.
5. An op amp has a (high, low) input impedance.
6. An op amp has a (high, low) output impedance.

FEEDBACK

Feedback is a technique used to modify the performance of an amplifier circuit. Positive feedback is generally used when increased gain or oscillation is desired. Negative feedback is commonly used to improve stability and linearity in amplifier circuits.

Negative Feedback

One way that an op amp is used in digital converters is in the closed-loop, or *negative feedback*, mode of operation. An op amp is usually operated in this mode, and the feedback is used to control the gain and stabilize the circuit. In the closed-loop configuration, the output signal is applied back to one of the input terminals through a resistor. The feedback signal always opposes the effects of the original input signal.

A negative feedback circuit is shown in Figure 9–4. For use in digital converters, the input with no signal is tied to ground. The input signal V_1 is applied through a resistor R_1 to the inverting terminal. The output signal is measured from the output terminal to ground. The output signal is also applied back through the feedback resistor R_f to the inverting input. The signal at the inverting input is determined not only by V_1 but also by some portion of V_{out}. The net result is that the amplifier "sees" a smaller input signal than is really present. The amplifier thus appears to have less gain than it did without the feedback.

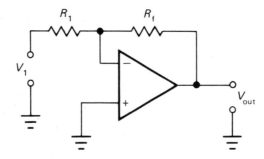

FIGURE 9–4

Negative Feedback Circuit

The resistors R_1 and R_f are selected to set the gain of the amplifier. The voltage gain of an op amp is expressed as follows:

$$A_v = -\frac{R_f}{R_1}$$

The minus sign indicates that the inverting input is being used. The voltage gain can be calculated by using the following equation:

$$A_v = -\frac{V_{out}}{V_1}$$

■ EXAMPLE

Look at Figure 9–5, which shows a typical circuit using negative feedback. The gain for this circuit is:

$$A_v = -\frac{R_f}{R_1} = -\frac{100 \text{ k}\Omega}{10 \text{ k}\Omega} = -10$$

If the input voltage $V_1 = 1$ V, the output voltage V_{out} will be:

$$V_{out} = V_1 \times -10 = 1 \times -10 = -10$$

The output will swing to a negative voltage only if the power supply inputs (not shown in the schematic) allow it to do so. In this example, $+V$ (pin 7) might be $+12$ V, and $-V$ (pin 4), -12 V.

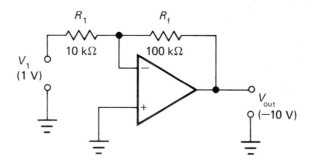

FIGURE 9–5

Inverting Amplifier

Positive Feedback

The gain for *positive feedback* is also determined by the resistors R_1 and R_f and is calculated as follows:

$$A_v = 1 + \frac{R_f}{R_1}$$

The gain will always be greater than 1. There is no minus sign here because the output is not inverted.

■ EXAMPLE

Look at Figure 9–6, which shows a typical circuit using positive feedback. The gain for this circuit is:

$$A_v = 1 + \frac{100 \text{ k}\Omega}{10 \text{ k}\Omega} = 1 + 10 = 11$$

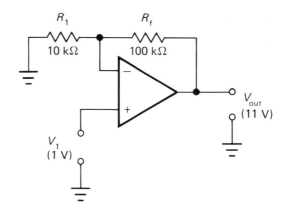

FIGURE 9–6

Noninverting Amplifier

SELF-TEST EXERCISE 9–2

1. Positive feedback results in (increased, decreased) gain.
2. Negative feedback is used to improve (gain, stability).
3. A negative feedback op amp circuit has $R_f = 10$ kΩ and $R_1 = 1$ kΩ. What is the gain of the circuit?
4. If the input voltage for the circuit in Question 3 is 1 V, what is the output voltage?
5. A positive feedback op amp circuit has $R_f = 20$ kΩ and $R_1 = 21$ kΩ. What is the gain of the circuit?
6. If the output voltage for the circuit in Question 5 is 16 V, what is the input voltage?

 STOP   Do Experiment 9–1

SUMMING AMPLIFIERS

Two or more signals are applied to the inverting terminal for a *summing amplifier*. The sum of these signals is taken from the output terminal, but the sum will be inverted. Figure 9–7 shows a summing amplifier circuit. This type of amplifier can be used to add dc or ac voltages. The formula for determining the output voltage is as follows:

$$V_{out} = -R_f \left(\frac{V_1}{R_1} + \frac{V_2}{R_2} + \cdots + \frac{V_z}{R_z} \right)$$

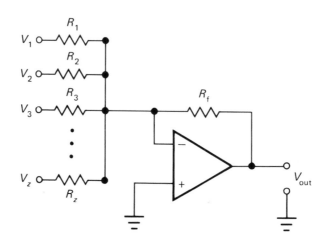

FIGURE 9–7

Summing Amplifier

where V_{out} = output voltage
R_f = feedback resistor
$V_1, V_2, \ldots V_z$ = input voltages
$R_1, R_2, \ldots R_z$ = input resistors

Any number of inputs can be connected in parallel. The size of R_f scales, or multiplies, the input levels by a constant. The gain of each individual channel is R_f/R. For converter circuits, the values of resistance are set so that each input has a different weight to represent the binary code.

COMPARATORS

Most op amp applications require a negative feedback. One exception occurs when an op amp is used as a *comparator*. The comparator is thus very important in our discussion of digital converters.

The comparator uses the full gain of the op amp (no feedback). The output reaches its saturation, or full-gain, level with very little input voltage. If an op amp has a gain of 200,000, it would take only a 0.1 mV input to result in a 20 V output. The output, however, cannot exceed the supply voltages since amplifiers do not originate power but just control available power. If we assume, in this case, that the op amp is being powered with $+10\,V$ as $+V_{CC}$ and ground as $-V_{CC}$, then the output cannot exceed 10 V. As a matter of fact, it will most likely be slightly less than 10 V because some of this voltage will be used for the internal operation. Therefore, V_{out} maximum can be only about $+9\,V$ maximum. The lowest V_{out} would be ground.

An op amp amplifies the difference between the two voltages at its input. If the voltage at the inverting input is even 0.1 mV more positive than that at the noninverting input, the output voltage will be driven to the most negative level—in this case, 0 V. If the voltage at the noninverting input is slightly more positive than the voltage at the inverting input, the output voltage will swing to the most positive level—in this case, almost 10 V.

The comparator operation is necessary in building an analog-to-digital converter. A comparator is always at one extreme or another; the output is never in-between. So, we can use the output as a signal to a digital circuit if we keep the voltage within the voltage limits for a binary one and zero.

SELF-TEST EXERCISE 9–3

1. What is V_{out} for the summing amplifier in Figure 9–8?

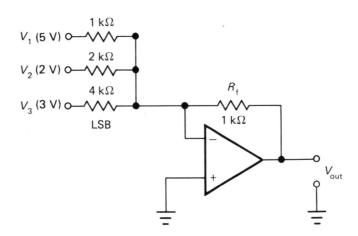

FIGURE 9–8

2. A comparator uses (the full gain, no gain) of the op amp.
3. If the negative input of an op amp is 1 V and the positive input is −1 V, what is the output of the comparator? The supply voltages are +5 V and −5 V.
4. For the comparator in Question 3, if the inverting input is −1 V and the noninverting input is 1 V, what is the output?

DIGITAL-TO-ANALOG CONVERTERS

A digital circuit can be used to control an analog device only if the digital value is converted to an analog equivalent value. Figure 9–9 shows the block diagram of a typical digital-to-analog converter. The heart of a converter is a *resistive network*, or *conversion ladder*. It provides a means of implementing a binary coding system that produces an output equivalent to some portion of a selected reference.

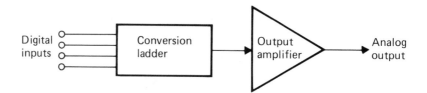

FIGURE 9–9

Block Diagram for Digital-to-Analog Converter

There are two types of DAC circuits: the binary-weighted resistor network or summing amplifier and the *R–2R* ladder network. These circuits produce an output that is an analog equivalent of the digital inputs but accomplish the conversion in different ways. In a *unipolar* DAC, for example, when all digital input lines are high, the analog output will be the full-scale output voltage. When all digital inputs are low, the output will be 0 V. In a unipolar DAC, the output voltage swings in only one direction.

Binary-Weighted DACs

For a *binary-weighted DAC*, a voltage is generated for each bit position in the binary number. The value of this voltage corresponds to the weight of the binary bit. For example, in the binary number 0001, the one bit represents the LSB position or a weight of 1. In the binary number 1000, the one bit represents the MSB position or a weight of 8. If the LSB location (binary 0001) has a weight of 1 V, then the binary number 1000 must have a value of 8 V. If the LSB location has a weight of 0.25 V, then the MSB of a 4-bit DAC must have a value of 2 V, or eight times the weight of the LSB. If more than one bit is high, as in the binary number 0110, then the resulting output must be the sum of the weights 2 and 4, or 6 V (for LSB = 1 V), which is proportional to the original binary number.

A summing amplifier multiples each input voltage by the ratio of the feedback resistor (R_f) to the corresponding input resistor (R). Therefore, for a given input voltage (V_{in}), we have the following general equation:

$$V_{out} = -V_{in} \left(\frac{R_f}{R} \right)$$

Remember, the minus sign indicates that the input to the op amp is applied to the inverting input; a plus sign indicates the noninverting input is being used.

The circuit in Figure 9–10 can be used to build a simple 4-bit resistor network DAC. The four switches can be electronic switches but are shown here as analog switches for simplicity. The four summing resistors are each selected with a weight proportional to the next. If a 10 kΩ MSB resistor is selected, then each progressive resistor must be twice the size of the preceding one. That is, R_d will be 10 kΩ, R_c will be $2R$ or 20 kΩ, R_b will be $4R$ or 40 kΩ, and R_a will be $8R$ or 80 kΩ. An op amp (U_1) and a feedback resistor (R_f) are selected to amplify the resulting sum of the inputs to a value needed for the control operation. A reference voltage must also be selected; in this case, it is $+10$ V.

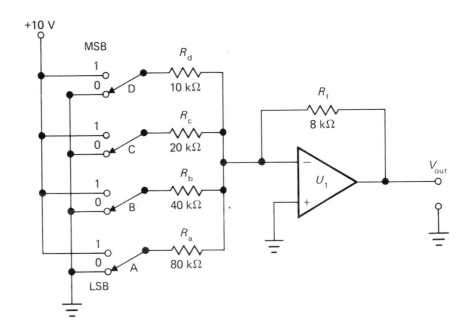

FIGURE 9–10

Binary-Weighted Resistor Network DAC

Notice that, as this circuit operates with all switches to ground, the input to the op amp is 0 V, and therefore the output is 0 V. Moving switch A to $+V_{ref}$ (binary 0001) will apply $+10$ V reference to the input of the op amp through the $8R$ or 80 kΩ resistor and will result in an output voltage of -1 V. [Remember the op amp formula $V_{out} = -V_{ref}(R_f/R)$]. In short, a 4-bit binary number represented by four switches has been converted into a voltage. It is the analog representation of one of the sixteen possible digital values. Moving switch B to V_{ref} will add another -2 V to the previous value of -1 V. The result is -3.0 V out. Table 9–1 shows all of the possible digital inputs and the corresponding output voltages for this circuit.

The resistors and voltages must be precise for this circuit to work as calculated. Therefore, actual DAC circuits contain precision amplifiers called *precision-level amplifiers* (PLAs) between the digital input and the corresponding input resistor to the summing amplifier. The precision reference supply feeds the level amplifiers, making the result a very precise output voltage of 0 V or V_{ref}.

DCBA	V_{out} Formula	V_{out} (V)
0000	$-V_{ref}\,(R_f/R)$	0
0001	$-V_{ref}\,(R_f/R_a)$	-1
0010	$-V_{ref}\,(R_f/R_b)$	-2
0011	$-V_{ref}\,(R_f/R_a + R_f/R_b)$	-3
0100	$-V_{ref}\,(R_f/R_c)$	-4
0101	$-V_{ref}\,(R_f/R_c + R_f/R_a)$	-5
0110	$-V_{ref}\,(R_f/R_c + R_f/R_b)$	-6
0111	$-V_{ref}\,(R_f/R_c + R_f/R_b + R_f/R_a)$	-7
1000	$-V_{ref}\,(R_f/R_d)$	-8
1001	$-V_{ref}\,(R_f/R_d + R_f/R_a)$	-9
1010	$-V_{ref}\,(R_f/R_d + R_f/R_b)$	-10
1011	$-V_{ref}\,(R_f/R_d + R_f/R_b + R_f/R_a)$	-11
1100	$-V_{ref}\,(R_f/R_d + R_f/R_c)$	-12
1101	$-V_{ref}\,(R_f/R_d + R_f/R_c + R_f/R_a)$	-13
1110	$-V_{ref}\,(R_f/R_d + R_f/R_c + R_f/R_b)$	-14
1111	$-V_{ref}\,(R_f/R_d + R_f/R_c + R_f/R_b + R_f/R_a)$	-15

R–2R Ladder DACs

An *R–2R ladder DAC* is useful where a large number of inputs are involved. Only two values of resistors (R and $2R$) are used in the ladder network, regardless of the number of stages. The feedback resistor (R_f) can be any value selected to provide the output scale level. If R_f is greater than R, the output is scaled up; if R_f is less than R, the output is scaled down.

In the circuit shown in Figure 9–11, the $R–2R$ ladder network is connected to the inverting input of the op amp. Each digital input is tied to the op amp adder circuit through a resistor that has a value of $2R$. The input closest to the op amp is the MSB. If a bit is zero, that switch is grounded; a one bit is switched to V_{ref}. The output voltage (V_{out}) is expressed by the following equation:

$$V_{out} = -\frac{V_{ref}}{2^N} \times \text{decimal value of binary input} \times \frac{R_f}{R}$$

where N = number of bits

For example, the output voltage for the circuit in Figure 9–11 is calculated as follows:

$$V_{out} = -\frac{5\,V}{2^4} \times 0110_2 \times \frac{10\,k\Omega}{50\,k\Omega} = -0.3125\,V \times 6 \times 0.2 = -0.375$$

If a particular output voltage were required, the digital input for that voltage would have to be calculated. For example, given an 8-bit DAC with

$$V_{ref} = +10\ V$$

$$\frac{R_f}{R} = \frac{10\ k\Omega}{10\ k\Omega}$$

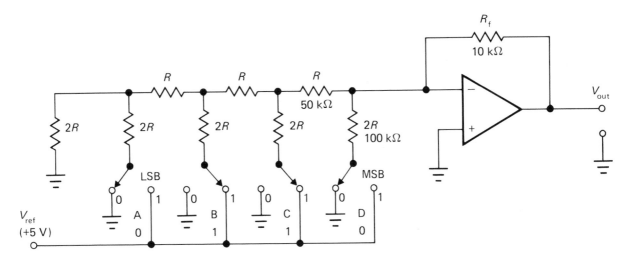

FIGURE 9–11

R–2R Ladder Network DAC

$$2^N = 2^8 = 256 \text{ steps}$$

$$\frac{10 \text{ V}}{256 \text{ steps}} = 39 \text{ mV per step}$$

$$V_{out} \text{ required} = 5 \text{ V}$$

then, the digital input would have to be:

$$\frac{V_{out} \text{ required}}{\text{volts per step}} \times \frac{R_f}{R} = \text{number of steps}$$

$$\frac{5 \text{ V}}{39.1 \text{ mV}} = 128 \text{ steps (rounded off)}$$

If the digital input is 128, the binary number applied to the digital inputs of the DAC would be 10000000.

Current DACs

Some DACs output current instead of voltage. The current output is a direct function of the digital inputs. Instead of a reference voltage, *current DACs* use a reference current. Reference current is set by the reference voltage and resistor. A reference voltage of +10 V would require a 10 kΩ resistor to set the reference current of the DAC to 1 mA. A variable resistor is used for the 10 kΩ to adjust for the exact current.

A current DAC can be connected to provide a voltage output from an external op amp. The whole circuit then becomes a voltage DAC. The output of the DAC will give an output current that is in the direction shown in Figure 9–12 and proportional to the digital outputs. The output current (I_{out}) is expressed by the following equation:

$$I_{out} = I_{ref} \left(\frac{1}{2^N} \times \text{decimal value of binary input} \right)$$

where N = number of bits

For example, if the eight digital inputs to the circuit in Figure 9–12 are set to 01100100, which equals 100 in decimal, then the output current is calculated as follows:

$$I_{out} = I_{ref} \left(\frac{1}{2^8} \times 100 \right) = 1 \text{ mA} \left(\frac{1}{256} \times 100 \right) = 0.39 \text{ mA}$$

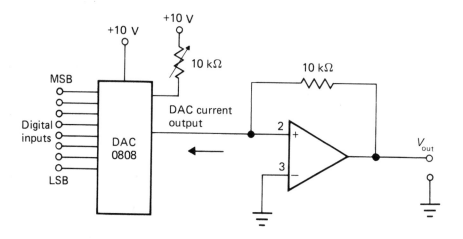

FIGURE 9–12

Current DAC Converted to Voltage DAC

DAC Resolution

DAC resolution is determined by the number of digital input lines available. These digital lines are used to control the scaling network. Resolution can be expressed in many different terms. Many manufacturers specify DAC resolution as the number of input bits. For example, if a DAC has ten digital input lines, it is referred to as a 10-bit DAC.

Resolution can also be expressed as the number of levels that can be produced at the output. If a DAC has 2^N output levels (where N is the number of digital inputs), then a 10-bit DAC has 2^{10} or 1024, output levels.

A better expression of resolution is the minimum amount of voltage that the output can change—that is, the LSB change from one step to the next. This step change is determined by dividing the maximum differential voltage at the output by the maximum number of unique states. For a 10-bit DAC with a reference voltage of +10 V, the minimum voltage change is 10 V/(2^{10}) = 10 V/1024 = 10 mV. The output voltage will change 10 mV for each LSB change of the digital inputs.

The percent resolution is calculated by the following equation:

$$\% \text{ resolution} = \frac{\text{step size}}{\text{FSR}} \times 100 \%$$

where step size = value of LSB
       FSR = full-scale voltage output

Another equation for determining the percent resolution is as follows:

$$\% \text{ resolution} = \frac{1}{\text{total number of steps}} \times 100\%$$

SELF-TEST EXERCISE 9–4

1. A unipolar DAC can swing in only one direction. True or False?
2. The LSB in an 8-bit DAC has a weight of 0.05 V. What is the weight of the MSB?
3. For the DAC in Question 2, what is the maximum analog voltage out?
4. For the digital-to-analog circuit in Figure 9–13, what is V_{out}?

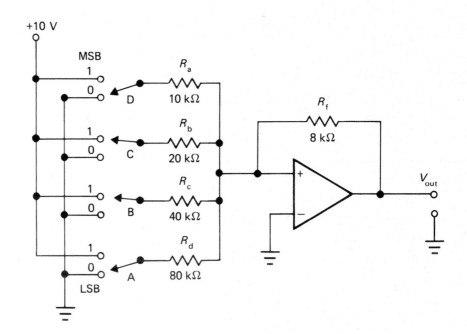

FIGURE 9–13

5. For the R–$2R$ ladder DAC in Figure 9–14, what is V_{out}?

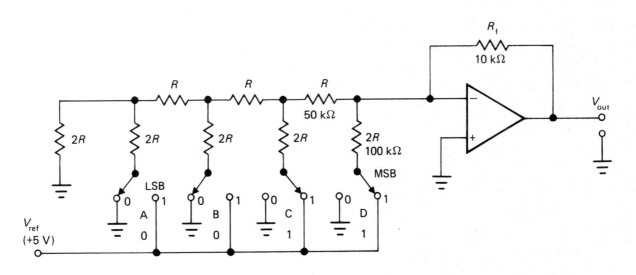

FIGURE 9–14

6. If a 4-bit current DAC has an I_{ref} of 2 mA, what is I_{out} for the digital input 0100?

7. The current DAC in Question 6 has a maximum output current of _____ .

STOP   Do Experiment 9–2

ANALOG-TO-DIGITAL CONVERTERS

The input to an analog-to-digital converter is an analog voltage; the output is a digital representation of that analog voltage. ADCs are used to input the digital equivalent of an external device's analog output voltage to a computer. The computer can make decisions based on the digital value of the voltage. ADCs are usually more expensive and complex than DACs because they use an internal DAC.

A *transducer* is a device that converts one form of energy to another. Transducers are used to convert the state of physical quantities such as temperature, liquid flow, pressure, light intensity, speed, strain, and vibration to electrical analog signals that can be used as inputs to the ADC.

In practical ADCs, the DAC output is used to supply a voltage to a comparison circuit. This circuit compares the voltage to be converted to the summed voltage from the DAC. The resultant output of the comparison circuit tells us whether the summed voltage is equal to the voltage to be converted. The comparator will switch logic states when these two voltages are the same. The digital value of the DAC is equivalent to the analog value at this point.

Typical digital output codes are BCD or binary. They can be positive or negative logic. The data sheet for an individual ADC indicates the type of output formatting and the conversion time. Usually, the faster the ADC is, the more it costs.

ADC Operation

Examining how a simple ADC operates will make the operation of other types easier to understand. Figure 9–15 shows the block diagram of a typical ADC circuit. This simple ADC contains a summing amplifier DAC. A discussion of the other elements shown in the block diagram will help us understand the operation of ADCs.

Analog Voltage Input.   The *analog voltage input* (V_a) is the voltage to be converted to a digital word. The acceptable limits for an analog voltage to the ADC without damaging the converter must be within these specifications.

A *bipolar* input voltage means that the analog voltage input may be positive or negative with respect to ground. A *unipolar* input voltage means that the input is either all positive or all negative with respect to ground.

Digital Output Word.   A *digital output word* is the resulting binary output after the conversion is complete.

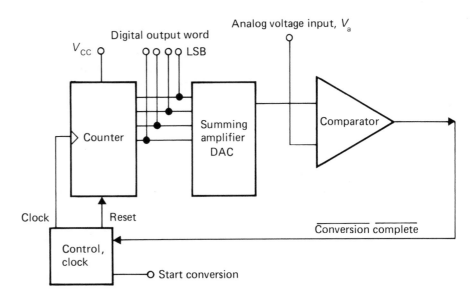

FIGURE 9–15

Block Diagram for Analog-to-Digital Converter

Start Conversion. The *start conversion* digital input informs the ADC that an analog voltage is stable and ready to be converted.

Convert Command. The *convert command* is the same as the start conversion. It is triggered on the positive or negative transition. There is a minimum and a maximum trigger time.

End Conversion. The *end conversion* output informs the external hardware that the analog voltage input has been converted and that the digital word is valid. The end conversion output is also called the *status* output.

Power Supply Inputs. The *power supply inputs* provide the supply voltages to operate all internal circuits in the ADC device.

ADC Resolution

ADC resolution is expressed as the maximum number of states of an ADC. If an ADC has 2^N states (where N is the number of bits) then a 12-bit ADC has a resolution of 2^{12}, or 4096, bits. The more states, the better the resolution.

If the maximum output voltage is 10 V, then the resolution in volts is expressed as the maximum output voltage divided by the number of resolution bits:

$$\frac{10 \text{ V}}{4096} = 2.4 \text{ mV per step}$$

So, for each step, the voltage will increase by 2.4 mV.

Digital Ramp ADCs

Digital ramp ADCs, also called *counting ADCs*, use the simplest and slowest conversion method but the least amount of hardware. Many commercial devices exist that contain several ICs, a DAC, a voltage comparator, a counter, and so on, all in a single package.

Figure 9–16 shows the block diagram of a typical digital ramp ADC circuit. The operation of this ADC is as follows. The V_a line is the analog voltage to be converted. The output of the comparator (status output) is initially high because the DAC output is not equal to the applied analog voltage V_a. A start pulse resets the counter to zero while inhibiting the AND gate. When the AND gate is inhibited, no clock pulses can pass to the counter. The DAC output voltage starts at 0 V. When start goes low, the AND gate is enabled. The clock pulses are allowed to pass to the counter. The clock pulses thus determine the output of the counter. Each clock pulse increases the counter output by one, which causes the DAC output to increase or increment toward the full-scale output voltage by the amount of voltage associated with the LSB in equal steps.

When the DAC output reaches a value greater than or equal to V_{in}, the comparator output goes low, placing a binary zero on the AND gate and turning it off. The conversion complete signal indicates when this data should be read. At this moment, we know that the DAC input

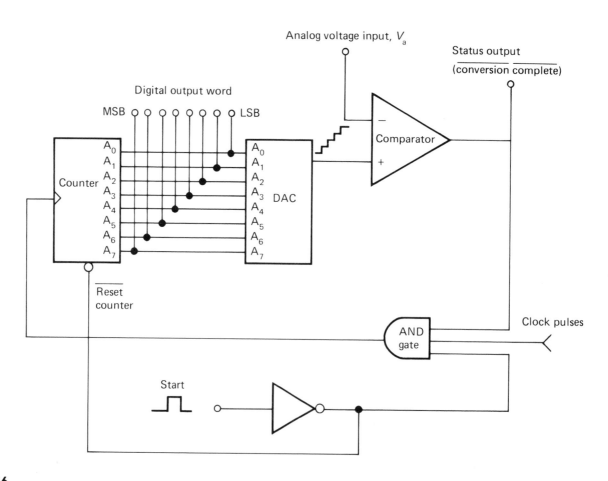

FIGURE 9–16

Block Diagram for Digital Ramp ADC

is equal to the analog input. The ADC output then reflects the digital value of V_a.

The time necessary for the conversion is determined by the frequency of the clock pulse. This frequency is limited to the speed of the counter and the DAC response.

For the circuit in Figure 9–16, suppose that the voltage to be converted is equal to 3.6 V and the comparator output is initially high. The digital inputs (through the counter) are incremented starting from 00000000. When the digital input is 00000001, the DAC outputs the analog value of 00000001. The DAC in the previous section increments in 39 mV steps; therefore, the output of the DAC is 39 mV. Since this output is less than 3.6 V, the counter continues to increment the DAC until the digital input 01011101 is applied. At this point, the DAC output is equal to 3.627 V. The comparator now changes state. This signal indicates the conversion is complete and inhibits the counter. The output of the ADC can now be read.

Successive-Approximation ADCs

Successive-approximation ADCs are the most widely used ADCs. They go through a sequence of approximations to obtain the digital representation of the analog input voltage. The maximum conversion time is much shorter than it is for counter-type ADCs and is a fixed value independent of the value of V_a. However, more complex circuitry is required.

Figure 9–17 shows the block diagram of a typical successive-approximation ADC circuit. It uses a register instead of a counter to provide input for the DAC. The binary values of the register are operated on in the following manner:

1. A binary number is entered into the register with the MSB set to one and all other bits at zero. The value at the DAC output is equal to

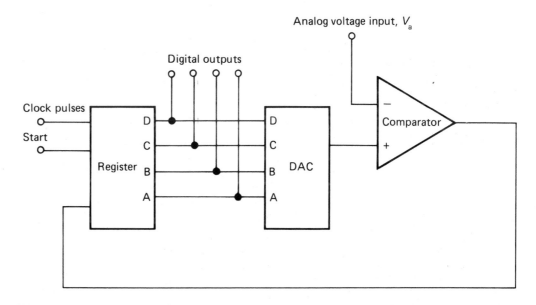

FIGURE 9–17

Block Diagram for Successive-Approximation ADC

the weight of the MSB. If this value is greater than V_a, the comparator goes low, resetting the MSB to zero. If the value is less than V_a, the MSB stays at one.

2. The next significant bit is set to one, and the comparison is again made as in the first step. If the new value is greater than V_a, the bit is reset to zero; otherwise, it remains at one.

3. This sequence continues for all bits in the binary number of the register. The process requires one clock pulse per bit.

4. After all bits have been tested, the clock pulses stop. The register now contains the digital equivalent of the analog input.

The following examples show that the conversion time of the successive-approximation ADC is much faster than that of the digital ramp ADC.

■ EXAMPLE

Given an 8-bit digital ramp ADC with a 1 MHz clock, the maximum conversion time (t) is:

$$t = 2^N \times \frac{1}{F} = 2^8 \times 1 \ \mu s = 256 \ \mu s$$

where N = number of bits   ■

■ EXAMPLE

Given an 8-bit successive-approximation ADC with a 1 MHz clock, the maximum conversion time (t) is:

$$t = N \times \frac{1}{F} = 8 \times 1 \ \mu s = 8 \ \mu s$$

where N = number of bits   ■

SELF-TEST EXERCISE 9–5

1. The input to an ADC is (analog, digital).
2. A transducer converts one form of energy to another. True or False?
3. A 10-bit ADC has a V_{ref} of 10 V. What is the LSB resolution?
4. What is the digital output for the digital ramp ADC in Figure 9–18 when the conversion is complete?
5. What was the conversion time for the ADC in Question 4?
6. What is the digital output for the successive-approximation ADC in Figure 9–19 when the conversion is complete?
7. What was the conversion time for the ADC in Question 6?

STOP   Do Experiment 9–3

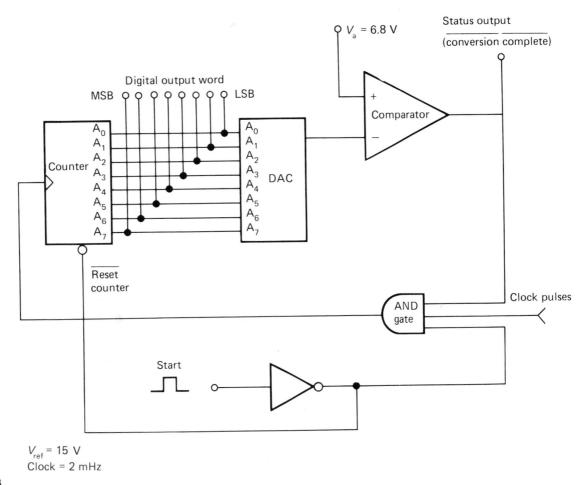

V_{ref} = 15 V
Clock = 2 mHz

FIGURE 9–18

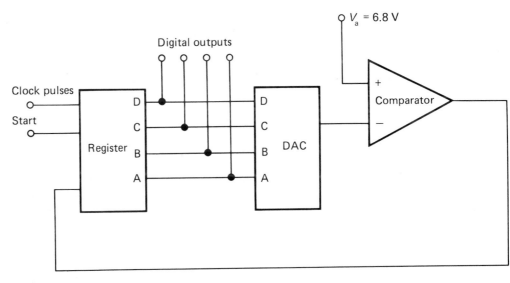

V_{ref} = 10 V
Clock = 2 mHz

FIGURE 9–19

SUMMARY

The operational amplifier, or op amp, is a special type of high-gain amplifier. The output of this circuit is tied back to one of the inputs to affect the total gain of the circuit. Positive feedback results in increased gain; negative feedback, in decreased gain but increased stability. The formula for determining the gain with negative feedback is:

$$A_v = -\frac{R_f}{R_1}$$

The formula for determining the gain with positive feedback is:

$$A_v = 1 + \frac{R_f}{R_1}$$

One application for the op amp is in digital converters as a comparator. The comparator is very sensitive to any difference in the two input voltages because it uses the full gain of the op amp (no feedback).

Digital-to-analog converters (DACs) are used to convert a binary input to an analog output. DACs make digital or computer control of analog devices possible. Computers cannot communicate with "outside world" devices (industrial equipment, home appliances, and so forth) without them.

Analog-to-digital converters (ADCs) are used to tell digital devices or a computer what the outside world is doing. This information can then be used to alter an operation if necessary.

CHAPTER 9
REVIEW EXERCISES

1. If negative feedback decreases the gain of an op amp, why is it used?

2. A negative feedback op amp circuit has R_f = 12 kΩ and R_1 = 2.2 kΩ. What is the gain of the circuit?

3. If the input voltage for the circuit in Question 2 is 0.5 V, what is the output voltage?

4. A positive feedback circuit has R_f = 22 kΩ and R_1 = 2.2 kΩ. What is the gain of the circuit?

5. If the output voltage for the circuit in Question 4 is 18 V, what is the input voltage?

6. What is V_{out} for the summing amplifier in Figure 9–20?

7. If the negative input of an op amp is 0.5 V and the positive input is -0.55 V, what is the output of the comparator? The supply voltages are $+10$ V and -10 V.

8. The LSB in a 12-bit DAC is 0.0125 V. What is the weight of the MSB?

9. What is the maximum analog output voltage for the DAC in Question 8?

10. What is V_{out} for the digital-to-analog circuit in Figure 9–21?

11. An 8-bit current DAC has an I_{ref} of 2 mA. What is I_{out} for the digital input 00011000?

12. The current DAC in Question 11 has a maximum output current of
_____.

13. An 8-bit ADC has a V_{ref} of 12 V. What is the LSB resolution?

14. What is the digital output for the digital ramp ADC in Figure 9–22 when the conversion is complete?

15. What was the conversion time for the ADC in Question 14?

16. What is the digital output for the successive-approximation ADC in Figure 9–23 when the conversion is complete?

17. What was the conversion time for the ADC in Question 16?

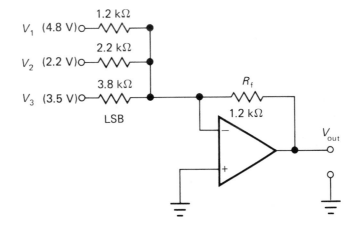

FIGURE 9–20

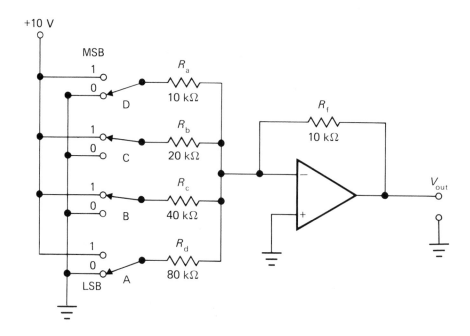

FIGURE 9–21

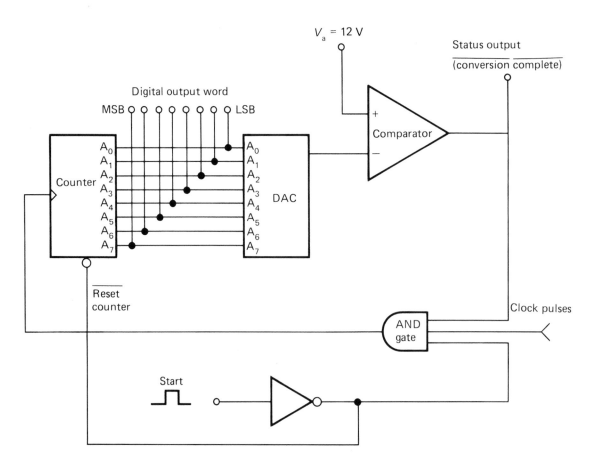

V_{ref} = 15 V
Clock = 10 MHz

FIGURE 9–22

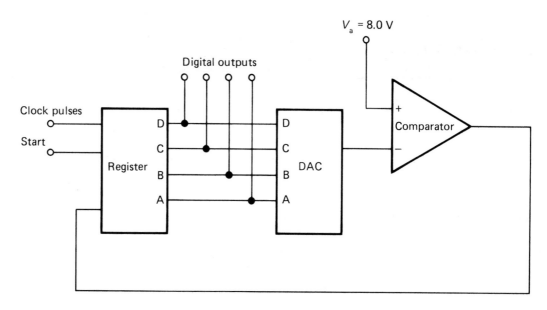

$V_{ref} = 10$ V
Clock = 5 MHz

FIGURE 9–23

EXPERIMENT 9–1 | Operational Amplifiers

PURPOSE

This experiment is designed to show the operation of operational amplifiers. You will build and test a negative and a positive feedback amplifier.

EQUIPMENT

—2 10 kΩ resistors (R_1, R_{f1})
—1 100 kΩ resistor (R_{f2})
—1 741 op amp (U_1)
—1 10 kΩ potentiometer (R_p)
—1 digital experimenter (Equip$_1$)
—1 oscilloscope or digital VOM (Equip$_2$)

PROCEDURE: PART I

Step 1 Build the circuit as shown in Figure 9E1–1.

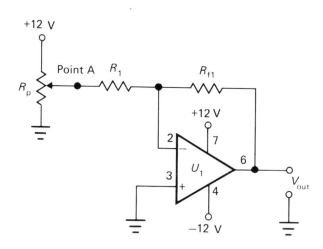

FIGURE 9E1–1

Step 2 Apply power to the circuit.

Step 3 Set the voltage at point A to 1 V by turning the potentiometer.

ACTIVITY

Measure the output voltage.

NOTE: The output should be −1 V because the gain of this circuit is −1: $-(R_{f1}/R_1) = -(10\ \text{k}\Omega/10\ \text{k}\Omega) = -1$.

Step 4 Turn off the power.

Step 5 Replace R_{f1} with R_{f2}.

Step 6 Turn on the power.

ACTIVITY

Measure the output voltage.

NOTE: The gain for this circuit is -10: $-(R_{f2}/R_1) = -(100 \text{ k}\Omega/10 \text{ k}\Omega)$ $= -10$. If the input voltage is 1 V, the output voltage will be -10 V.

Step 7 Turn off the power. Replace R_{f2} with R_{f1}.

Step 8 Turn on the power.

ACTIVITY

Measure the output voltage while you slowly increase the input voltage at point A. Record this information in the table in Figure 9E1–2.

NOTE: The maximum output voltage will not reach -12 V.

V_{in} (point A)	V_{out}
+1 V	
+2 V	
+3 V	
+4 V	
+5 V	
+6 V	
+7 V	
+8 V	
+9 V	
+10 V	
+11 V	
+12 V	

FIGURE 9E1–2

PROCEDURE: PART II

Step 1 Build the circuit as shown in Figure 9E1–3.

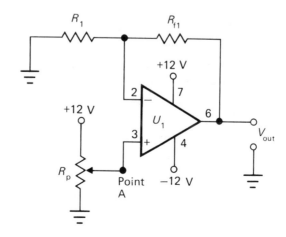

FIGURE 9E1–3

Step 2 Apply power to the circuit.

Step 3 Set the voltage at point A to 1 V by turning the potentiometer.

ACTIVITY

Measure the output voltage.

NOTE: The output should be +2 V because the gain of this circuit is 2: $(R_{f1}/R_1) + 1 = (10 \text{ k}\Omega/10 \text{ k}\Omega) + 1 = 2$.

Step 4 Turn off the power.

Step 5 Replace R_{f1} with R_{f2}.

Step 6 Turn on the power.

ACTIVITY

Measure the output voltage.

NOTE: The gain for this circuit is 11: $(R_{f2}/R_1) + 1 = (100 \text{ k}\Omega/10 \text{ k}\Omega) + 1 = 11$. If the input voltage is 1 V, the output voltage will be 11 V.

Step 7 Turn off the power. Replace R_{f2} with R_{f1}.

Step 8 Turn on the power.

ACTIVITY

Measure the output voltage while you slowly increase the input voltage at point A. Record this information in the table in Figure 9E1-4.

NOTE: The maximum output voltage will be less than 12 V.

V_{in} (point A)	V_{out}
+1 V	
+2 V	
+3 V	
+4 V	
+5 V	
+6 V	
+7 V	
+8 V	
+9 V	
+10 V	
+11 V	
+12 V	

FIGURE 9E1-4

EXPERIMENT 9–2 | Digital-to-Analog Converters

PURPOSE

This experiment is designed to show the operation of DACs. You will build and test a binary-weighted and an R–$2R$ ladder DAC.

EQUIPMENT: PART I

—1 100 kΩ 1% resistor (R_d)
—1 200 kΩ 1% resistor (R_c)
—1 400 kΩ 1% resistor (R_b)
—1 800 kΩ 1% resistor (R_a)
—1 80 kΩ 1% resistor (R_f)
—1 741 op amp (U_1)
—1 digital experimenter ($Equip_1$)
—1 oscilloscope or digital VOM ($Equip_2$)

PROCEDURE: PART I

Step 1 Build the circuit as shown in Figure 9E2–1.

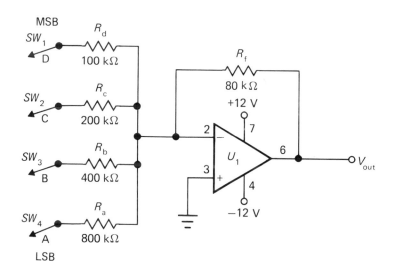

FIGURE 9E2–1

Step 2 Apply power to the circuit.

ACTIVITY

Complete the table in Figure 9E2–2 by using the data switches for the binary input.

NOTE: You may adjust the calculated values by using the measured values of V_{ref}, R_f, and R. The accuracy of your measurements will be determined largely by the tolerance of the resistors you have selected for the circuit. The closer they are to the specified value, the better your test results will be.

DCBA	V_{out} formula	Calculated	Measured
0000	$-V_{ref}\,(R_f/R)$	0 V	
0001	$-V_{ref}\,(R_f/R_a)$	0.5 V	
0010	$-V_{ref}\,(R_f/R_b)$	1.0 V	
0011	$-V_{ref}\,(R_f/R_a + R_f/R_b)$	1.5 V	
0100	$-V_{ref}\,(R_f/R_c)$	2.0 V	
0101	$-V_{ref}\,(R_f/R_c + R_f/R_a)$	2.5 V	
0110	$-V_{ref}\,(R_f/R_c + R_f/R_b)$	3.0 V	
0111	$-V_{ref}\,(R_f/R_c + R_f/R_b + R_f/R_a)$	3.5 V	
1000	$-V_{ref}\,(R_f/R_d)$	4.0 V	
1001	$-V_{ref}\,(R_f/R_d + R_f/R_a)$	4.5 V	
1010	$-V_{ref}\,(R_f/R_d + R_f/R_b)$	5.0 V	
1011	$-V_{ref}\,(R_f/R_d + R_f/R_b + R_f/R_a)$	5.5 V	
1100	$-V_{ref}\,(R_f/R_d + R_f/R_c)$	6.0 V	
1101	$-V_{ref}\,(R_f/R_d + R_f/R_c + R_f/R_a)$	6.5 V	
1110	$-V_{ref}\,(R_f/R_d + R_f/R_c + R_f/R_b)$	7.0 V	
1111	$-V_{ref}\,(R_f/R_d + R_f/R_c + R_f/R_b + R_f/R_a)$	7.5 V	

FIGURE 9E2–2

EQUIPMENT: PART II

—4 50 kΩ 1% resistors (R, R_f)
—5 100 kΩ 1% resistors $(2R)$
—1 741 op amp (U_1)
—1 digital experimenter $(Equip_1)$
—1 oscilloscope or digital VOM $(Equip_2)$

PROCEDURE: PART II

Step 1 Build the circuit as shown in Figure 9E2–3.

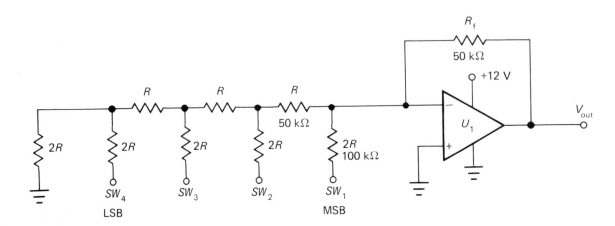

FIGURE 9E2–3

Step 2 Apply power to the circuit.

ACTIVITY

Complete the table in Figure 9E2–4 by using the data switches for the binary input.

NOTE: The calculated values may be adjusted by using the actual (measured) values of R_f and R. Don't forget any variation in V_{ref}.

DCBA	V_{out}	
	Calculated	Measured
0000	0	
0001	0.3125	
0010	0.625	
0011	0.9375	
0100	1.25	
0101	1.5625	
0110	1.875	
0111	2.1875	
1000	2.5	
1001	2.8125	
1010	3.125	
1011	3.4375	
1100	3.75	
1101	4.0625	
1110	4.375	
1111	4.6875	

FIGURE 9E2–4

LSB = 0.3125

ACTIVITY

Compare the results of Part I and Part II.

NOTE: The R–$2R$ circuit should be more accurate than the binary-weighted circuit, providing care was taken to select the resistors and accurate measurements were taken.

EXPERIMENT 9–3

Analog-to-Digital Converters

PURPOSE

This experiment is designed to show the operation of ADCs. You will build and test an 8-bit ADC.

EQUIPMENT

—1 ADC 0808 ADC IC (U_1)
—8 330 Ω resistors (R_1–R_8)
—1 10 kΩ potentiometer (R_p)
—8 LEDs (any color) (L_1–L_8)
—1 digital experimenter (Equip$_1$)
—1 digital voltmeter (Equip$_2$)

PROCEDURE

NOTE: The ADC 0808 is an 8-bit ADC, 8-channel multiplexer. Since only one channel is used for this experiment, the three address lines are tied to ground to use channel zero. Channel zero is input IN$_0$ (pin 26).

Five volts is used for both V_{CC} and V_{ref} to make it easier to use the digital experimenter. If you have access to an experimenter with eight or more logic indicators and variable dc voltages, the experiment can be done a second time with a different V_{CC} and V_{ref}. The eight outputs Q_0–Q_7 and the V_{ref} determine the resolution of this circuit. The LSB is 5 V/2^8, or approximately 0.0195 V.

Step 1 Build the circuit as shown in Figure 9E3–1.

NOTE: Use the LEDs (L_1–L_8) for the digital output indicators. The end of conversion (EOC) signal at pin 7 can be tied to one of the logic indicators on the experimenter. Don't forget the current-limiting resistors for L_1–L_8. The 10 kΩ potentiometer is used to create a varied analog input voltage from 0 to +5 V.

ACTIVITY

Carefully set the analog input at point A to the voltages in the table in Figure 9E3–2. Use the digital voltmeter to measure these as accurately as possible. When the input is set and stable, start the conversion by depressing $\overline{B}$ once. Depress $\overline{A}$ until the EOC indicator lights to show the conversion has been completed. Record the binary output indicated by L_1–L_8 in the table.

NOTE: The experimenter clock can be used instead of $\overline{A}$ to speed things up, of course, but try to do the first few with the logic switch.

This experiment should give very good results if you have carefully wired the circuit. If the 10 kΩ potentiometer you are using is a multi-turn type, you will have even better control of the input voltage and can make many more measurements since the analog input can be varied more carefully.

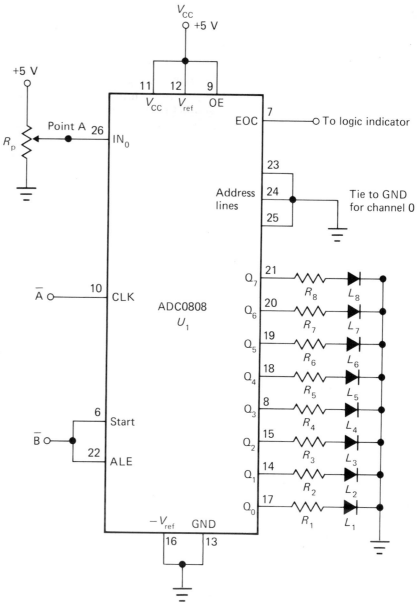

FIGURE 9E3–1

Approximate decimal value		.0195	.039	.078	.1563	.3125	.625	1.25	2.5
V_{in}		Q_7	Q_6	Q_5	Q_4	Q_3	Q_2	Q_1	Q_0
0 V									
+0.2 V									
+0.5 V									
+1.0 V									
+1.5 V									
+2.0 V									
+2.5 V									
+3.0 V									
+3.5 V									
+4.0 V									
+4.5 V									
+5.0 V									

FIGURE 9E3–2

This experiment has used very little of the capabilities of this IC. Each of the inputs IN_0–IN_7 can be used. A different device could be connected to each input and these inputs monitored at intervals. A counter applied to the address lines selects the input to be monitored. The speed of this counter determines the intervals. The clock input to the ADC must be faster than the counter so that time is allowed for the conversion.

This IC is easily interfaced to a computer. The computer can then monitor eight devices accurately in very little time.

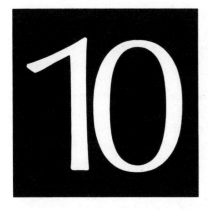

Semiconductor Memory Circuits

OBJECTIVES

After studying this chapter, you will be able to:

1. Define a semiconductor memory.
2. Define the terms used to describe memory types, characteristics, and functions.
3. Describe the difference between sequential-access memory and random-access memory.
4. Give the characteristics of MOS memory and bipolar memory.
5. Build a simple RWM circuit. Write sixteen 4-bit words into the memory. Read sixteen 4-bit words from memory.
6. Define the four types of ROM described in this chapter.
7. Describe some features and applications of RWM.
8. Describe some features and applications of ROM.
9. List the steps for troubleshooting ROMs and RWMs.
10. Name the three states for tri-state TTL.

INTRODUCTION

Digital circuits become more useful when they can remember what to do and when to do it. Memory circuits remember what they are programmed to remember. Some of these circuits are used to store data for short periods of time, while others contain programs that actually "run" computers.

In digital electronics, *memory* is any device that can store a logic one or a logic zero in a manner such that a single bit or group of bits (word) can be accessed or retrieved as required. The main use for memory is in computer circuits.

Memory devices are divided into two main categories of memory: read/write memory (RWM) and read-only memory (ROM). ROM and RWM are LSI devices. RWM is used for short-term storage; ROM is used for permanent storage of data. This chapter describes the operation of these memory components. To help you understand their operation, the terms, characteristics, types, and functions related to these semiconductor memories are defined first.

MEMORY TERMS

Write.   To *write* means to enter a data word into a memory location.

Read.   To *read* means to retrieve a data word from a memory location.

Access Time.   The *access time* is the time it takes to read one word out of memory or to write one word into memory. Some modern memory devices have access times that are less than 75 ns (nanoseconds). This time is sometimes referred to as the propagation delay.

Data.   The *data* is the information or binary word to be stored.

Address.   The *address* is the specific memory location for data. It designates where the information or data is stored. Each word of data is written into or read from memory using only this address.

Capacity.   The *capacity* of a memory chip is specified in terms of the maximum number of bits or the maximum number of words that the memory can store.

MEMORY CHARACTERISTICS

Memory devices have some or all of the following characteristics. These characteristics determine not only the memory application but also the cost.

Random-Access Memory.   In *random-access memory*, any location in memory can be addressed without numerically sorting through all the locations before the specified one is reached. An example of how random-access memory works is a phonograph record; any position on the record can be touched by the tone arm to start playing.

Sequential-Access (Serial-Access) Memory.   In *sequential-access memory*, data words are read one after another in physical sequence. A magnetic (cassette) tape is an example of how sequential-access memory works. Reading a particular word requires first passing through all the words in front of it. The time it takes to access a word depends on how far the selected address is from the beginning of the tape. Memories composed of shift registers are another example of sequential-access memory. (This chapter covers only random-access memories.)

Destructive and Nondestructive Readout.   In *destructive readout*, a data word is lost as it is read. In *nondestructive readout*, a data word remains in memory even after it is read.

Static and Dynamic MOS Memory.   In *static memory*, data enters as a charge, logic, or magnetic state that remains in a memory cell until a new word is written. A set/reset flip-flop is an example of how static memory works; a set flip-flop stays set until it is reset or until the circuit loses power.

There is no bipolar counterpart for dynamic MOS memory, which is a high-capacity, moderate-speed, low power consumption memory. In *dynamic memory*, data must constantly be rewritten. Dynamic memory is built with a charge-storage capacitor in series with a MOS driver transistor. The absence or presence of a charge in the capacitor is interpreted by the RWM sense line as a zero or one. This capacitor must constantly be recharged to retain its state. Recharging, or *refreshing*, is done about every 2 ms (milliseconds). Constantly rewriting the data causes the memory to appear constant. Once refreshing in dynamic memory is stopped, the data is lost. Timing is critical for refreshing dynamic memory components. Some dynamic RWMs have complex refresh circuitry on the chips.

Both static memory and dynamic memory have advantages and disadvantages. One advantage of static RWM is that it is faster than dynamic RWM is. Static memory can have an access time of only 20 ns, while dynamic memory access time may be as long as 100 ns. However, static memory does not have as many memory cells (storage space) as common dynamic memory does. Dynamic memory is much less expensive than static memory is but requires more support circuitry for the refresh operation. Finally, dynamic memory is more susceptible to noise than static memory is, and noise can cause "soft-fail" (errors in data).

Both static memory and dynamic memory may be found in the same system. Static memory is usually used in systems less than or equal to 64K, while dynamic memory is often used in systems greater than or equal to 64K.

Volatile and Nonvolatile Memory. *Nonvolatile memory* is memory that does not lose its contents when power is removed from the circuit. All ROMs are nonvolatile. Core memory is an example of nonvolatile memory.

Volatile memory is memory that does not retain data during a power failure. RWM, whether static or dynamic, depends on circuit power to avoid loss of data. All RWMs are, therefore, volatile. To temporarily make an RWM nonvolatile, battery backup is used in some equipment. It is usually used with dynamic MOS memory. Note that, if an RWM is called nonvolatile by the manufacturer, it is a hybrid type that also contains some type of ROM. This hybrid type is known as NVRAM, for nonvolatile random-access memory.

**SELF-TEST
EXERCISE 10–1**

1. _____ is sometimes referred to as the propagation delay.
2. Destructive readout means the RWM _____ when it is read.
3. To (read, write) into RWM is to store data in the memory cells.
4. To (read, write) from RWM is to have information appear on the data output lines.
5. Which is faster: random-access memory or sequential-access memory?
6. _____ RWM must be refreshed so that data will not be lost.
7. The location where data is stored is called the (data, address).

MEMORY TYPES

Bipolar Memory. *Bipolar memory* is faster, more powerful, and easier to interface than MOS memory is. It is, however, more expensive and takes more space for the same amount of storage. Bipolar memory is used in applications where speed is required.

MOS Memory. The three major MOS technology families—*PMOS*, *NMOS*, and *CMOS*—are classified by the channel type of the MOS transistors in the chip. PMOS memory has positive-charged carriers; NMOS has negative-charged carriers; and CMOS is a combination of both. NMOS memory is the fastest because of the higher speed of electrons in silicon. CMOS is usually used in battery-operated or battery-backup applications. MOS memory has a high density and low power dissipation, although it is slower than TTL. CMOS memory uses less power than NMOS does. NMOS memory is faster and denser than PMOS is. NMOS memory is usually TTL compatible because it uses only one power supply voltage.

MOS GUIDELINES

Caution is required when MOS devices are to be handled or tested. To avoid possible static damage to these ICs, manufacturer-recommended standard procedures should include the following guidelines:

1. Store MOS devices in conductive foam so that all leads are shorted together.
2. Make sure that the person handling the MOS device is grounded.
3. Use grounded conductive mats over nonconducting surfaces at all work stations.
4. Connect all conductive surfaces and equipment to earth ground.
5. Wear rubber gloves and clothing that does not generate static when you are handling MOS parts.
6. Handle all MOS parts by their packages and not by their leads.

Core Memory. In *core memory*, each bit of data is stored in a small, doughnut-shaped, permanent magnet. Core memory is a large, expensive, and high power consumption memory.

Magnetic Bubble Memory. *Magnetic bubble memory* (MBM) is a magnetic storage technique somewhat similar to core memory but is smaller in size and consumes less power. MBM is nonvolatile and physically very rugged. It is used most often in severe environments or remote sites. It is also used for some portable terminals. Bubble memory has exceptional storage capacity.

ECL Memory. *ECL memory* is the fastest and most powerful type of memory. It is, however, the most expensive.

READ/WRITE MEMORY (RWM)

Read/write memory (RWM) is a semiconductor memory in which data can be written into (stored) and read out (retrieved). Any location in RWM is accessible without regard to any other location; it is thus a random-access memory (RAM). Although both RWM and ROM are randomly accessible, RWM is still often called by the generic term RAM to the exclusion of ROM. RWM is used in computer systems for temporary storage of data. It holds data until processing is completed or until power is lost. Since RWM devices lose their memory when power is turned off, RWM systems either must be left on all the time, or battery backup must be added, or important data must be stored on a nonvolatile medium. All RWMs except NVRAMs are volatile unless some type of power backup is supplied.

Figure 10–1 shows the pin assignments for an Intel 2164A dynamic RWM and an Intel 2114A static RWM. Notice the difference in the control pins. A discussion of the inputs and outputs of a typical RWM follows.

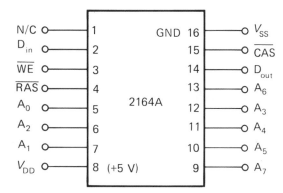

A. 65K X 1 Dynamic RWM

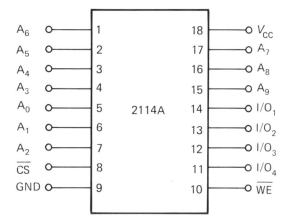

FIGURE 10–1

Dynamic and Static RWM Pin
Assignments

B. 1K X 4 Static RWM

RWM Inputs and Outputs

1. The memory cell select input or address input selects the specific memory cell or group of memory cells (location) desired for the read/write operation.

2. The write enable ($\overline{\text{WE}}$) or read/write ($\text{R}/\overline{\text{W}}$) pin on the chip determines whether the RWM is to write into memory or to output the data in memory. When $\overline{\text{WE}}$ or $\text{R}/\overline{\text{W}}$ is low, the chip is in the write mode; when $\overline{\text{WE}}$ or $\text{R}/\overline{\text{W}}$ is high, it is in the read mode.

3. The data inputs are the lines where data are entered for storage when a read is commanded. Some RWMs, such as the Intel 2114A, have only one set of data lines that is used for input and output depending on the state of the $\text{R}/\overline{\text{W}}$ control.

4. The chip select ($\overline{\text{CS}}$) pin allows selection of an individual IC when memory is tied to common bus lines.

5. The column address strobe ($\overline{\text{CAS}}$) lines are used only with dynamic RWMs.

6. The row address strobe ($\overline{\text{RAS}}$) lines are used with dynamic RWMs. $\overline{\text{RAS}}$ is used for the refresh operation to refresh one row of memory at a time. Sequencing through all the row addresses at least every 2 ms keeps the memory refreshed.

7. The data outputs are the outputs that provide the data lines out for the information stored in RWM. RWM devices are available with

static storage and nondestructive readout and are erasable. They are all normally volatile.

RWM Configuration

RWMs of different storage capacities are available. They can be configured in many different ways. For example, a 64×4 RWM is one that has 64 4-bit words. The total bit capacity is 256 bits (64×4). A $1K \times 4$ RWM has 1024 4-bit words (K is equal to 1024 bits). Its total capacity is 4096 bits.

The total number of address inputs depends on the number of addresses available in the memory array. In general, a memory requires N address lines, where 2^N is the number of addresses in the array. An address selects a single bit in a 1-bit–organized memory chip and selects an entire word in a word-organized memory chip. The length of the word is determined by the number of data lines. A 64×4 RWM must have six address lines ($2^6 = 64$) to address all 64 locations. It needs four data lines for the 4-bit word.

To select the proper cell or word in the memory array, a binary word is placed on the address line and indicates where the data is to go or come from in the chip. Some examples of common RAM configurations are as follows:

$$64 \text{ bits} = 16 \times 4 = 16 \text{ 4-bit words}$$

$$256 \text{ bits} = 256 \times 1 = 256 \text{ 1-bit words}$$

$$1024 \text{ bits} = 256 \times 4 = 256 \text{ 4-bit words}$$

$$1024 \text{ bits} = 1024 \times 1 = 1024 \text{ 1-bit words}$$

$$4096 \text{ bits} = 1024 \times 4 = 1024 \text{ 4-bit words}$$

$$4096 \text{ bits} = 4096 \times 1 = 4096 \text{ 1-bit words}$$

SELF-TEST EXERCISE 10–2

1. RWM is (volatile, nonvolatile) memory.
2. RWM can be made nonvolatile by using:
   a. battery backup
   b. refresh
   c. ECL logic
   d. TTL type
   e. none of the above
3. Core memory is (volatile, nonvolatile) memory.
4. Data lines serve the same purpose as address lines. True or False?
5. RWM is (random-access, sequential-access) memory.
6. What is the total bit capacity for a $4K \times 1$ RWM?
7. What is the total bit capacity for an $8K \times 8$ RWM?
8. How many data words does a $1K \times 4$ RWM have?
9. How many data input lines does a 256×4 RWM have?

STOP Do Experiment 10–1

READ-ONLY MEMORY (ROM)

Read-only memory (ROM) is a semiconductor randomly accessible memory from which digital data can repeatedly be read out but cannot be written in. During manufacture, once desired data is programmed into a ROM, there is no way that it can be altered (changed) without destroying the entire IC. There are, however, ROMs that are *erasable* or *reprogrammable* by special techniques.

The main difference between RWM and ROM is that RWM loses data when power is lost but ROM does not. ROM is, therefore, nonvolatile. ROM usually involves MOS rather than bipolar technology because more memory elements can be packed in the same amount of space. MOS technology also consumes less power.

ROM Types

Mask-Programmed ROM.   A *mask-programmed ROM* is permanently fixed by the manufacturer according to customer specifications. It is only used in high-volume production because the setup procedure is expensive. Once a particular mask is in production, however, it is the least expensive type of ROM and offers the most storage space.

PROM.   A *programmable read-only memory* (PROM) is programmed by the customer. It is a one-shot operation; once it has been programmed or fixed, it cannot be changed. If a change is required, a new PROM must be programmed.

To program a *fusible-link* PROM, the desired location for a word is addressed and the unwanted fuses are "blasted" out with a high voltage. The PROM is then permanently altered. If a mistake is made, the PROM must be thrown away and a new one blasted. Since a fusible-link PROM can easily be programmed by the customer at places other than the semiconductor factory, it is often referred to as a *field-programmable* ROM. Figure 10–2 shows the internal configuration of a fusible-link PROM.

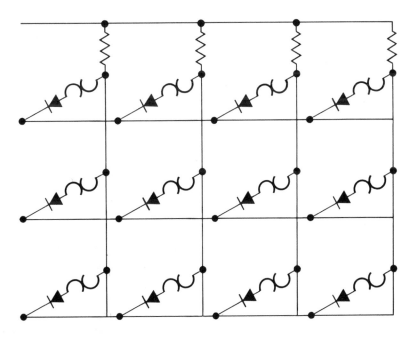

FIGURE 10–2

Internal Configuration of a
Fusible-Link PROM

EPROM. An *erasable programmable read-only memory* (EPROM) is programmed and erased by the user. The EPROM is an expensive device that must be programmed and erased by external devices. It is programmed by applying high-voltage pulses (about four to five times the normal supply voltage) to the gate of a transistor in the chip. The gate is then changed to a high impedance state. To erase the program, an ultraviolet light is projected through the quartz window in the top of the EPROM.

EAROM. An *electrically alterable read-only memory* (EAROM) is programmed and altered electrically. It does not need an outside device for programming. Special electrical signals applied for a few milliseconds are used to write into an EAROM. It is not as permanent as an EPROM since it usually can only be written into 10,000 times, but it is nonvolatile. Currently, the EAROM is the most expensive and least dense type of ROM. EAROMs are used in some TV channel selectors. They are also used for storage of temporary setup information for video terminals.

EEPROM. An *electrically erasable programmable read-only memory* (EEPROM) is programmed in the same way as an EAROM. Some of the newer versions of EEPROMs do not require any external support chips for this programming since they have circuitry in the chip to amplify the +5 V supply voltage to the programming voltage.

NVRAM. The new *nonvolatile random-access memory* (NVRAM) is a memory device that combines the qualities of ROM and RAM. It serves as a volatile memory chip until it receives a store signal. Data in memory is then promptly moved to a nonvolatile section of the IC.

**SELF-TEST
EXERCISE 10–3**

1. ROM is (random-access, sequential-access) memory.
2. Programs in ROM are generally (erasable, not erasable).
3. (ROMs, PROMs) are programmable by the user.
4. EPROMs can be erased by _____.
5. EAROMs are _____.
6. ROM is (nonvolatile, volatile) memory.
7. RWM and ROM are different in that only RWM has (read/write, address, data) lines.
8. ROM is randomly accessible memory. True or False?
9. (ROMs, RWMs) cannot be written into during normal use.

ROM Applications

Every system has some ROM circuitry because it must have at least enough built-in programming to load its RWM. The program contained in a ROM to start up a system is called the *bootstrap program*. This program "boots up" the system so that operation can begin.

Basically, ROMs are used when permanent or near-permanent memory is required or when a system always executes the same programs—that is, when information is not subject to change. ROMs can also be used to hold tables that are needed for computer calculations. Multiplication, division, sine, and cosine tables are common applications.

ROMs can substitute for combinational logic networks. For example, a ROM can be programmed to generate a complex truth table and thus avoid the use of a large number of gates or the high cost of a custom IC. The EPROM is an extremely powerful, efficient, and economical technique for generating complex circuits.

Comparison of RWM and ROM Operation

Figure 10–3 shows the block diagram for a TTL LSI 7489 16×4 RWM. This typical RWM consists of an address decoder, data buffer, memory cell array, and output buffer circuits. The address decoder is similar to any binary decoder. It accepts a multibit input word and decodes it. Only one of the decoder outputs will be enabled, which, in turn, selects the proper memory location. The RWM in Figure 10–3 has a 4-bit address line; therefore, sixteen different addresses can be decoded. Each of the sixteen addresses is a 4-bit word (note the number of data lines here).

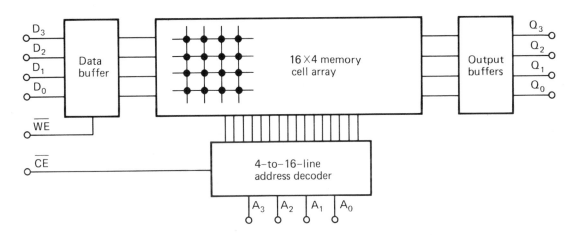

FIGURE 10–3

**Block Diagram for a TTL LSI
7489 16 × 4 RWM**

The main body of the memory, the memory cell array, consists of circuits that are used to store binary data. The type of circuits are dependent on the type of RWM or ROM. The storage elements are arranged in a way such that a specific number of multibit words can be stored. Applying a 4-bit address will cause the addressed location to accept or output the data (depending on the read/write line). Remember, ROM can only read (output) data. The output is sent to other circuits through the buffers. These buffers help prevent noise that may be on the bus lines from reaching the memory cells.

If two ROMs or RWMs are needed to increase the memory capability, the input lines are not doubled. To double memory requires only one additional address line. See Figure 10–4. Note that, if there are four address lines (16 words), only five address lines are needed for two memory ICs (32 words); six lines would address three memory ICs (48 words). In Figure 10–4, U_1 would be enabled for 0 to 15 (16 words), and U_2 would be enabled for 16 to 31 (another 16 words).

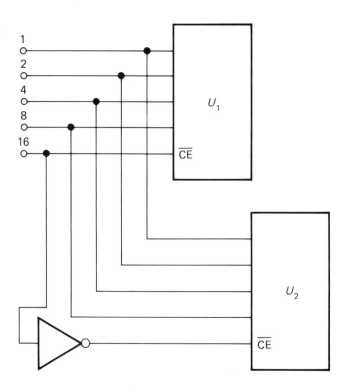

FIGURE 10–4

Doubling Memory Capability

Like any other circuit, ROMs have propagation delay. This delay, or access time, is the time between the application of an input address and the appearance of data at the outputs.

Like RWMs, ROMs are available in 16×8, 256×1, 64×8, 128×4 and many other configurations. Their organization can be changed by using latches or data selectors. For example, if a ROM has an 8-bit output and only 4-bit words are needed, a data selector can be used that will pick either of the four bits. See Figure 10–5. The number of words available is thus doubled. In the same way, a 16-bit word can be output by connecting two 8-bit latches at the outputs, as shown in Figure 10–6. In this case, the number of words available is reduced by one-half.

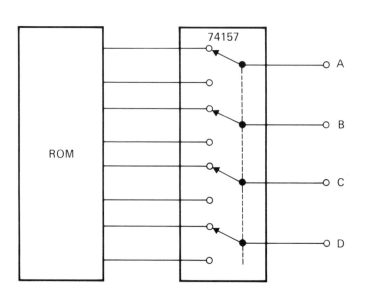

FIGURE 10–5

Doubling the Number of Words (8 Bits Changed to Two 4-Bit Words)

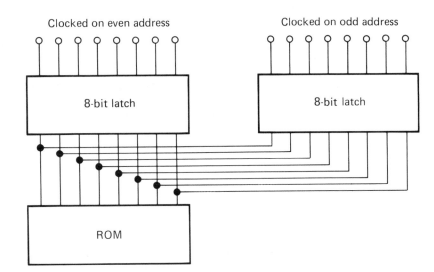

FIGURE 10-6

Reducing the Number of Words (Two 8-Bit Words Changed to 16 Bits)

SELF-TEST EXERCISE 10-4

1. How many memory locations can an RWM with six address lines decode?
2. How many individual memory cells does a 256×4 RWM have?
3. ROMs have no propagation delay. True or False?
4. One RWM is added to double the memory of an 8-line address RWM. How many additional address lines are needed?
5. A data _____ would be used if only half of the output bits are needed.

TROUBLESHOOTING ROMs AND RWMs

Many problems in memory circuits can be traced by using careful trouble-shooting methods. Never "jump into" a circuit. That is, take time to evaluate what is or is not happening. First check the power supply; then proceed with an in-depth examination.

1. Always check the power supply and ground lines. Memory ICs are very sensitive to poorly filtered input power. They are also easily damaged by low or high voltages. A very small negative voltage can permanently damage some ICs and RWMs in particular.
2. Check that each address line is pulsing at the correct voltage levels, V_{cc} and ground. Suspect any line that stays high or low or pulses at other than the V_{cc} and ground levels. Remove the ROM or RWM (if it is a plug-in type) and check the address line again. If the line pulses, the ROM or RWM is defective and must be replaced. If the address line does not pulse with the ROM or RWM removed, then the problem may be in several RWMs or ROMs tied to a common bus line. In this case, remove and replace all RWMs or ROMs one by one and discard any that cause a malfunction. It is not uncommon for all the memory chips to be damaged if a power supply problem exists. Check the previous stage of the circuit if the address lines do not pulse with all of the memory chips removed. Follow the procedure in step 6 if the problem still cannot be found.
3. Check that each data line is pulsing. Check each line as you did in step 2.

4. Be sure to check any enable lines. ICs that are not enabled do not function.

5. Check for an analog signal on any line. Such a signal indicates a defective RWM.

6. Use a *checkerboard test* if necessary. Input data at all RWM addresses and then check each data line to see whether or not the data was stored and can be retrieved. Use a "checkerboard" input, in which the first address is ones and zeros alternately, the second address is zeros and ones alternately, and so on. See Figure 10–7. Or, alternate all zeros or all ones for each successive address (for an 8-bit word, the data in the first address would be 00000000, the data in the second address would be 11111111, and so on) until all addresses are full. See Figure 10–8. Then read the information one address at a time. Shorted lines are easily detected by this method.

Address			Data in			
C	B	A	4	3	2	1
0	0	0	1	0	1	0
0	0	1	0	1	0	1
0	1	0	1	0	1	0
0	1	1	0	1	0	1
1	0	0	1	0	1	0
1	0	1	0	1	0	1
1	1	0	1	0	1	0
1	1	1	0	1	0	1

FIGURE 10–7

Checkerboard Test

Address			Data in			
C	B	A	4	3	2	1
0	0	0	0	0	0	0
0	0	1	1	1	1	1
0	1	0	0	0	0	0
0	1	1	1	1	1	1
1	0	0	0	0	0	0
1	0	1	1	1	1	1
1	1	0	0	0	0	0
1	1	1	1	1	1	1

FIGURE 10–8

Alternate Data Test

Note: Occasionally, the RWM has a *bit-sensitive error*, which means that the RWM is not able to store a specific binary word. In this case, a very lengthy test of all possible bit patterns is necessary using a microprocessor or specialized test equipment.

7. Check dynamic RWMs for any problems in the refresh circuit. Timing is critical if the RWMs are to be properly refreshed.

8. Always substitute an RWM suspected of having a defect with one that is known to be functional, if one is available.

TRI-STATE TTL ICs

All of the ICs that have been discussed thus far were assumed to have two possible output states, either *low* or *high*. Some ICs have another output state, called a *high impedance state*. Like the normal zero (low) and one (high) states, this third state is like an open switch. It represents a very high impedance and is equivalent to disconnecting the output circuit from the output pin of the IC. This type of output is very useful in multiplexed, or bused, transmission circuits.

A *bus* is a group of wires, lines, or cables over which binary information is transferred from one place to another. There are two types of buses: unidirectional and bidirectional. In *unidirectional* bus lines, data travels in only one direction. In *bidirectional* bus lines, data travels in both directions. Most data buses are bidirectional; most address buses are unidirectional. Instead of having many lines, one set of lines is time-shared by two points. When one source is transmitting, the other waits in a high impedance state until the transfer is complete. This concept of having one set of lines (a bus) that carries multiple signals is called *multiplexing*. The circuit that is not in use is disabled while the sending or receiving circuits are activated.

A bus line can be achieved by paralleling TTL open-collector ICs. Standard TTL ICs cannot be paralleled because of their totem-pole output connections. They must be externally wired with resistors. Therefore, a tri-state logic circuit was developed using basic TTL circuits. A control line was added to obtain the third state. The state of the control line determines whether the output is an open or is allowed to operate as a standard TTL. When the control line is low, the output is a standard TTL; when the control line goes high, the output goes to the high impedance (open) state. This state is sometimes shown on a truth table as the Z state since Z is the common symbol for impedance.

Figure 10–9 shows the logic symbol for a tri-state logic gate that can be paralleled to form a common bus line. When data is to be transmitted, all nontransmitting gates will have their control lines high. Only the gate designated to transmit data will be enabled.

Figure 10–10 shows a simple diagram of a typical bidirectional bus line using tri-state devices. Only one of several bus lines is shown. This bus line is responsible for transmitting one bit of information in either direction. Only one gate will be enabled at a time to transmit data. One bit of data can be transmitted in either direction from any one source to one or more destinations.

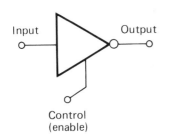

FIGURE 10–9

Logic Symbol for Tri-State Logic Gate

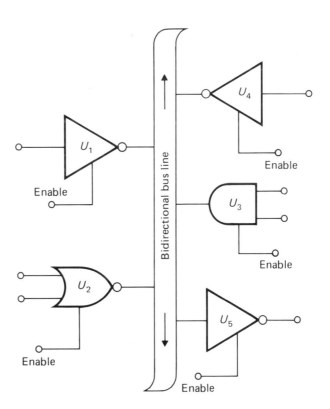

FIGURE 10–10

Bidirectional Bus Line Using Tri-State Devices

SELF-TEST EXERCISE 10–5

1. The first step in memory testing is to check the _____ to the ICs.
2. Data and _____ lines ordinarily pulse between V_{CC} and ground.
3. A ROM that is suspected of being defective can be removed and replaced with any other ROM. True or False?
4. Tri-state logic has three possible output states. List them.
5. Tri-state gates can be paralleled to form _____ lines.
6. Standard TTL ICs (can, cannot) be paralleled to form bus lines.

 **STOP** Do Experiment 10–2

SUMMARY

A semiconductor memory is any device that can store a logic one or a logic zero. In this book, we have discussed many types of digital circuits that can be defined as temporary memory—for example, the JK flip-flop, the RS latch, and decoder circuits. This chapter dealt with devices that are used for the defined purpose of memory.

Read/write memory (RWM), also commonly called RAM, is a semiconductor memory in which data can be randomly accessed, read and written, and lost when power is lost (volatile). Read-only memory (ROM) is the memory device that is used for more permanent data storage. In ROM, data can be randomly accessed, can normally only be read, and is not lost when power is lost (nonvolatile).

Destructive memory is memory that is lost when it is read. ROM is nondestructive memory, in which a data word remains in memory even after it is read. RWM can be either destructive or nondestructive. In a static memory cell, data entry causes a change of state in the cell. This change is retained until a new word is written or until power is lost. In dynamic memory, data must continually be rewritten or it will be lost.

Bipolar memory is easier to use and faster but takes more space than MOS memory for the same amount of storage. MOS memory uses less power but must be handled carefully. The type of memory selected depends on which features are most important for an application. Core memory, magnetic bubble memory, and ECL memory have some exceptional qualities but are more expensive than bipolar or MOS memories.

A mask-programmed ROM is programmed during manufacture. Some special ROMs are programmed by the user. A fusible-link PROM can be programmed only once by the user. An EPROM is expensive but can be programmed many times by the user in a special programming device. An EPROM is erased by shining an ultraviolet light through the quartz window in the top of the EPROM. EPROMs are used most commonly in engineering design or manufacturing operations that require occasional program modifications or changes.

EAROMs and EEPROMs are the most expensive type of ROM, but they are also the easiest to use because they can be programmed in the circuit. Currently, they cannot be reliably programmed as many times as EPROMs. They simply "wear out" after 10,000 to 1,000,000 write cycles, which is not enough time for most field use. EEPROMs are electrically erasable as well as electrically programmable. EAROMs can be electrically reprogrammed.

NVRAMs combine the qualities of RAM and ROM. Part of a NVRAM IC acts as volatile memory, while a special section is used for nonvolatile reprogrammable memory. NVRAMs have a short life expectancy.

A variety of methods can be used to extend memory storage capability. The number and size of data words stored can be increased by the addition of decoding ICs.

CHAPTER 10
REVIEW EXERCISES

1. Define a semiconductor memory.

2. Which of the following features are applicable to an RWM?

   a. It stores data permanently.
   b. Data can be read and written.
   c. It has data input lines, address lines, and an R/$\overline{\text{W}}$ line.
   d. It can be accessed sequentially only.
   e. All of the above apply.

3. An RWM has eight address lines. How many locations can be addressed?

   a. 128
   b. 8
   c. 256
   d. 255
   e. 64

4. What is destructive readout?

5. A ROM has an 8-bit address line and a 4-bit output data line. What is the memory capability?

   a. stores eight 4-bit words
   b. stores four 8-bit words
   c. stores sixteen 8-bit words
   d. stores 256 4-bit words
   e. stores four 256-bit words

6. State the main difference between ROM and RWM.

7. What is meant by the term *write*?

8. Define access time.

9. A 7489 is a (RWM, RAM). What is the memory capability for the 7489?

10. What is an EAROM?

11. Give an example of a ROM application.

12. What is the first step in troubleshooting any IC?

13. List the three states of a tri-state TTL.

14. How are EPROMs erased?

15. What is the difference between static and dynamic RWM?

EXPERIMENT 10–1 | 64×4 RWM Circuit

PURPOSE

This experiment is designed to demonstrate the function of an RWM circuit. You will store binary information at the specified addresses and then read the information.

EQUIPMENT

—1 digital experimenter (Equip₁)
—1 logic probe (Equip₂)
—1 7489 64×4 RWM (U_1)
—4 1 kΩ, 1/4 W resistors (R_1–R_4)

PROCEDURE

NOTE: The 7489 is a 64×4 RWM. Figure 10E1–1 shows the pin assignment for this IC, which has four data lines in and four data lines out. The four address lines allow access to 64 separate locations, each one four bits long. This RWM can thus read or write 64 4-bit words.

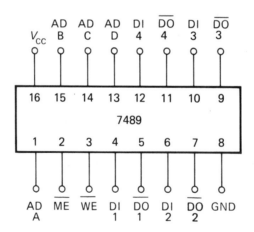

FIGURE 10E1–1

The truth table for the 7489 is shown in Figure 10E1–2. In mode 1, the IC outputs the $\overline{stored\ data}$. The line over *stored data* indicates that the output is the complement (inverse) of the actual data word that was stored. The memory enable ($\overline{ME}$) must be low to access the chip. The write enable ($\overline{WE}$) is high for the read operation. Any information on the data input (IN) lines is ignored.

In mode 2, to write a zero in the RWM, the data input (IN) line must be low. The write enable ($\overline{WE}$) must be low to indicate the write function. The memory enable ($\overline{ME}$) is then taken low to enable the chip. The complement of the data input is seen on the data output line ($\overline{DN}$).

In mode 3, to write a one in the RWM, the data input (IN) line must be high. The write enable ($\overline{WE}$) must be low to indicate the write

Mode	Chip enable ($\overline{ME}$)	Read/write ($\overline{WE}$)	Data in	Data out ($\overline{DN}$)
1 (Read)	0	1	X	Stored data
2 (Write 0)	0	0	1	1
3 (Write 1)	0	0	0	0
4 (Inhibit)	1	X	X	High

FIGURE 10E1–2

function. The memory enable ($\overline{ME}$) is then taken low to enable the chip. The complement of the data input is seen on the data output line ($\overline{DN}$).

In mode 4, to disable the IC, the memory enable ($\overline{ME}$) is taken high. All other inputs are ignored. All data outputs go high.

Step 1 Construct the circuit as shown in Figure 10E1–3.

NOTE: The 1 kΩ resistors R_1–R_4 are used as pull-up resistors because the IC has open-collector outputs.

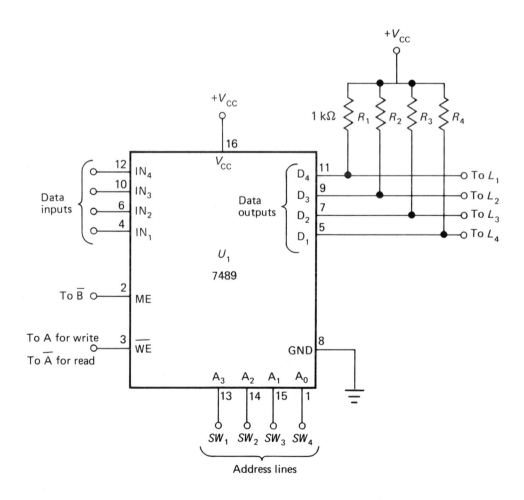

FIGURE 10E1–3

LAB

ACTIVITY

Sequence the address from 0000 to 1111 using switches SW_1–SW_4. This activity tells the IC where you want to store the data. Input the data words indicated in the table in Figure 10E1–4 by connecting the data input lines to ground or +5 V as shown.

NOTE: Remember that, after you set the address location and apply the proper data inputs, you must enable the IC by depressing logic switch $\overline{B}$ connected to the memory enable ($\overline{ME}$); otherwise, nothing will be stored. If you have correctly stored data at all locations, the information will remain in the RWM until power is lost.

Address (memory location)	Data (information in)	Decimal equivalent
0000	1000	8
0001	0010	2
0010	0101	5
0011	1001	9
0100	0011	3
0101	0111	7
0110	0110	6
0111	1000	8
1000	0111	7
1001	0011	3
1010	0001	1
1011	0010	2
1100	0100	4
1101	0101	5
1110	0111	7
1111	1001	9

FIGURE 10E1-4

ACTIVITY

After all data has been stored, read the data from each location. Move logic switch A to $\overline{A}$ to indicate to the IC that you want to read. Select the proper address using the switches; then depress $\overline{B}$ to enable the IC. The information stored at each location will be displayed at logic indicators L_1–L_4.

NOTE: The displayed results are the complement (inverse) of the actual input data. The table in Figure 10E1–5 shows the information as it will be displayed.

Address	Data
0000	0111
0001	1101
0010	1010
0011	0110
0100	1100
0101	1000
0110	1001
0111	0111
1000	1000
1001	1100
1010	1110
1011	1101
1100	1011
1101	1010
1110	1000
1111	0110

FIGURE 10E1–5

<table>
<tr><td>**EXPERIMENT 10–2**</td><td># Tri-State Buffer</td></tr>
</table>

PURPOSE

This experiment is designed to demonstrate the operation of an inverting tri-state buffer. You will connect a tri-state buffer to the RWM circuit in Experiment 10–1.

EQUIPMENT

—1 digital experimenter (Equip$_1$)
—1 logic probe (Equip$_2$)
—1 digital multimeter (Equip$_3$)
—1 7489 64×4 RWM (U_1)
—1 74368A inverted tri-state output bus driver (U_2)
—4 1 kΩ resistors (R_1–R_4)

PROCEDURE

Step 1 Construct the circuit as shown in Figure 10E2–1.

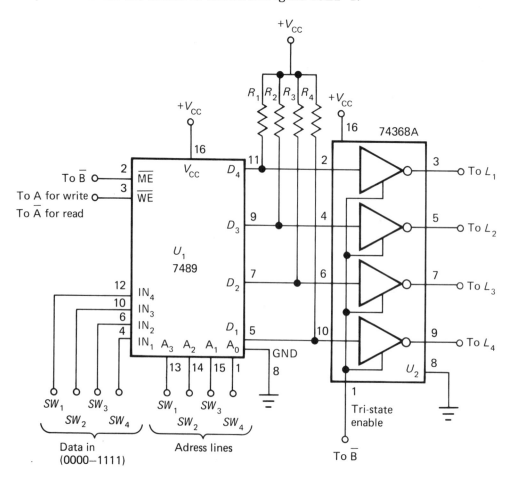

FIGURE 10E2–1

ACTIVITY

Store the data in the table in Figure 10E2–2 as you did in Experiment 10–1.

NOTE: To make it easier to input the data for this experiment, the binary numbers 0000 through 1111 are used. Therefore, the address locations are the same as the data to be stored in the IC. Don't be confused, however. Remember from Experiment 10–1 that anything can be stored at each location.

Address				Data inputs			
D	C	B	A	4	3	2	1
0	0	0	0	0	0	0	0
0	0	0	1	0	0	0	1
0	0	1	0	0	0	1	0
0	0	1	1	0	0	1	1
0	0	0	0	0	1	0	0
0	0	0	1	0	1	0	1
0	0	1	0	0	1	1	0
0	0	1	1	0	1	1	1
1	1	0	0	1	0	0	0
1	1	0	1	1	0	0	1
1	1	1	0	1	0	1	0
1	1	1	1	1	0	1	1
1	1	0	0	1	1	0	0
1	1	0	1	1	1	0	1
1	1	1	0	1	1	1	0
1	1	1	1	1	1	1	1

FIGURE 10E2–2

NOTE: This tri-state buffer is an inverting buffer. Its truth table is shown in Figure 10E2–3. All data inputs will be displayed at logic indicators L_1–L_4 in the same state as the data in. Since the tri-state buffer only allows a signal to pass to the output when it is enabled, L_1–L_4 will only display data when the RWM and tri-state buffer are enabled with logic switch $\overline{B}$.

Input		Output
$\overline{E}$	D	
L	L	H
L	H	L
H	X	Z

FIGURE 10E2–3

ACTIVITY

Confirm that all of the data has been stored as you entered it by moving logic switch A to $\overline{A}$. Read each location and compare the output with the information you stored in Figure 10E2–2.

ACTIVITY

Now, with the tri-state buffer in the inhibit mode (the enable line high), use the logic probe to measure the output at pins 3, 5, 7, and 9. Record the results.

NOTE: Various logic probes give different indications for the high impedance state. Most probes indicate the high impedance state by giving no light. You should be familiar with how the probe you are using operates.

ACTIVITY

Use the digital meter to measure the voltage at the output pins. Before you make this measurement, disconnect logic indicators L_1–L_4 since they will affect your readings.

NOTE: You should have measured 2.2 V to 2.4 V. Do not touch the leads while you measure this high impedance state since doing so will affect the measurements. As you know, 2.2 V to 2.4 V is an unacceptable logic state. High impedance outputs are normally in this range, however, when they are tested with a meter.

Answers

SELF-TEST EXERCISE 1–1

1. analog and digital **2.** analog, continuously; digital, in discrete steps
3. a. digital **b.** digital **c.** analog **d.** analog **4.** are available
at low cost and are extremely versatile, reliable, and small **5.** reduction
in size, weight, cost, and power consumption **6.** True **7. b.** digital

SELF-TEST EXERCISE 1–2

1. 2 **2.** two steps **3. a.** 1010_2 **b.** 100011_2 **c.** 1100100_2
d. 1101011.11_2 **4. a.** 27 **b.** 55 **c.** 127 **d.** 33.375
5. $2^6 - 1 = 63$ **6.** 64 **7.** nine digits **8.** Binary-coded decimal
is a decimal system with 0 to 9 digits represented in a 4-bit binary code.
9. a. $0001\ 0010\ 0110_{BCD}$ **b.** $0111\ 0010\ 0011_{BCD}$ **c.** $0001\ 0101_{BCD}$
d. 1001_{BCD} **10. a.** 87 **b.** 339 **c.** 564 **11. a.** 2105_8
b. 710_8 **c.** 135_8 **12. a.** 2595 **b.** 255 **c.** 3018
13. a. $1B0_{16}$ **b.** $15BC_{16}$ **c.** 62_{16} **14.** False **15.** True
16. True **17.** True **18.** False **19.** False **20.** False
21. False

SELF-TEST EXERCISE 1–3

1. components and electronic circuits **2.** cut-off state and conduction
3. bipolar and MOS field effect transistors **4.** The most positive voltage
level is defined as the binary 1 state. **5.** The least positive voltage level
is defined as the binary 1 state. **6. a.** negative logic **b.** positive logic
c. positive logic **d.** negative logic **7.** parallel and serial **8. a.**
and **b.** **9. a.** and **b.** **10. c.** gates

11. +5 V

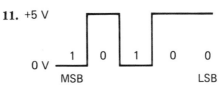

12. False **13.** False **14.** True **15.** True **16.** True
17. True **18.** False **19.** True **20.** True **21.** True
22. True **23.** True

SELF-TEST EXERCISE 2–1

1. AND, OR, inverter **2.** binary **3.** Memory or sequential logic circuits **4.** inverter **5.** decision-making or combinational
6. NOT A or $\overline{A}$ or the complement of A **7.** truth table **8.** False
9. inverter **10.** True **11.** static and dynamic **12.** 5 mA to
15 mA (All logic probes draw some current.) **13.** True
14. Dynamic **15.** False

SELF-TEST EXERCISE 2–2

1. decision-making **2.** is **3.** all

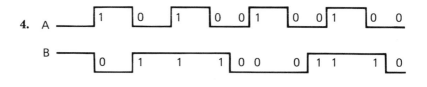

4. A

B

C

5. AND gate   **6.** 16 (because 2^4 = 16)   **7.** one output (multiple inputs)   **8.** ABC = X or A · B · C = X or A × B × C = X   **9.** +0.8 V to +2.4 V (a voltage level that is between the acceptable high or low state for TTL)   **10.** If A = 1, B = 0, and C = 0, then X = 1. If A = 1, B = 1, and C = 1, then X = 0.

SELF-TEST EXERCISE 2–3

1. decision-making   **2.** one or more   **3.** OR

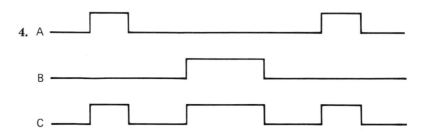

4. A

B

C

5. one   **6.** X + Y = Z   **7.** OR   **8.** X = (A · B) + C
9. a. AND   **b.** inverter   **c.** OR   **d.** OR

SELF-TEST EXERCISE 2–4

1 NAND   **2.** NOR   **3.** AND   **4.** NAND   **5.** inverter
6. Exclusive OR

7. False (Figure 2–42 is an Exclusive NOR symbol.)

8. AND   **9.** OR   **10.** False (This is equivalent to the inverter or NOT gate.)   **11. a.** NAND   **b.** OR   **c.** inverter   **d.** AND
e. NOR   **f.** Exclusive OR

SELF-TEST EXERCISE 2–5

1. True   **2.** high-power   **3.** Discrete   **4.** Integrated circuits
5. large-scale integration   **6.** False   **7.** Fan-out   **8.** ECL
9. RTL and DTL   **10.** False   **11.** more expensive   **12.** rise time (t_r)   **13.** +5 V   **14.** 5400   **15.** True   **16.** slower
17. True   **18.** CMOS   **19.** approximately 100

SELF-TEST EXERCISE 3–1

1. True   **2.** one   **3.** one   **4.** one   **5.** low   **6.** lows
7. high   **8.** highs   **9.** 0   **10.** True   **11.** low

SELF-TEST EXERCISE 3–2   **1.** temporary storage registers   **2.** high   **3.** low

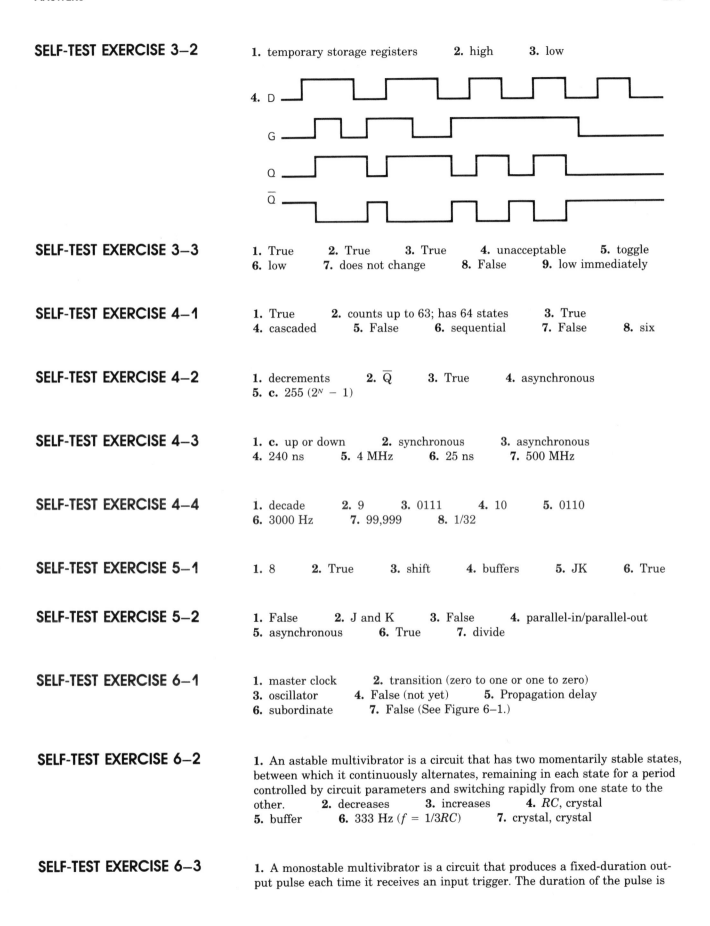

SELF-TEST EXERCISE 3–3   **1.** True   **2.** True   **3.** True   **4.** unacceptable   **5.** toggle
6. low   **7.** does not change   **8.** False   **9.** low immediately

SELF-TEST EXERCISE 4–1   **1.** True   **2.** counts up to 63; has 64 states   **3.** True
4. cascaded   **5.** False   **6.** sequential   **7.** False   **8.** six

SELF-TEST EXERCISE 4–2   **1.** decrements   **2.** $\overline{Q}$ **3.** True   **4.** asynchronous
5. c. 255 $(2^N - 1)$

SELF-TEST EXERCISE 4–3   **1. c.** up or down   **2.** synchronous   **3.** asynchronous
4. 240 ns   **5.** 4 MHz   **6.** 25 ns   **7.** 500 MHz

SELF-TEST EXERCISE 4–4   **1.** decade   **2.** 9   **3.** 0111   **4.** 10   **5.** 0110
6. 3000 Hz   **7.** 99,999   **8.** 1/32

SELF-TEST EXERCISE 5–1   **1.** 8   **2.** True   **3.** shift   **4.** buffers   **5.** JK   **6.** True

SELF-TEST EXERCISE 5–2   **1.** False   **2.** J and K   **3.** False   **4.** parallel-in/parallel-out
5. asynchronous   **6.** True   **7.** divide

SELF-TEST EXERCISE 6–1   **1.** master clock   **2.** transition (zero to one or one to zero)
3. oscillator   **4.** False (not yet)   **5.** Propagation delay
6. subordinate   **7.** False (See Figure 6–1.)

SELF-TEST EXERCISE 6–2   **1.** An astable multivibrator is a circuit that has two momentarily stable states, between which it continuously alternates, remaining in each state for a period controlled by circuit parameters and switching rapidly from one state to the other.   **2.** decreases   **3.** increases   **4.** RC, crystal
5. buffer   **6.** 333 Hz ($f = 1/3RC$)   **7.** crystal, crystal

SELF-TEST EXERCISE 6–3   **1.** A monostable multivibrator is a circuit that produces a fixed-duration output pulse each time it receives an input trigger. The duration of the pulse is

usually controlled by external components.   **2.** Increasing   **3.** 14.3 kΩ; use tables for the approximate value or calculate with $t_w = 0.7\,(C_{ext} \times R_t)$.
4. duty cycle   **5.** missing pulse or burst

SELF-TEST EXERCISE 6–4   **1.** False   **2.** True   **3.** will not   **4.** nanosecond   **5.** Two-phase clocks (or nonoverlapping clocks or delayed clocks)   **6.** t_H

SELF-TEST EXERCISE 7–1   **1.** encoder   **2.** keyboard   **3.** 001_2 **4.** 0110_2 (the active low logic level binary equivalent for 9)   **5.** largest

SELF-TEST EXERCISE 7–2   **1.** binary   **2.** an octal   **3.** True   **4.** 48 or 110000_2
5. current-limiting resistor   **6.** segments a, b, e, d, and g   **7.** BCD

SELF-TEST EXERCISE 7–3   **1.** multiplexer   **2.** demultiplexer   **3.** demultiplexer
4. multiplexer   **5.** data selector   **6.** multiplexer
7. demultiplexer   **8.** multiplexing   **9.** buses or bus lines

SELF-TEST EXERCISE 8–1   **1. b.** when only one input is binary one

2.
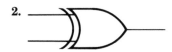

3.

A	B	C
0	0	0
0	1	1
1	0	1
1	1	0

4. Exclusive NOR gate   **5.** Exclusive NOR gate

6.

A	B	C
0	0	1
0	1	0
1	0	0
1	1	1

7.
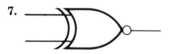

SELF-TEST EXERCISE 8–2   **1.** Exclusive OR gate   **2.** MSB or LSB   **3.** even   **4.** odd
5. a. 1   **b.** 0   **c.** 0   **d.** 0   **e.** 1   **6. a.** 0   **b.** 1   **c.** 1
d. 1   **e.** 0   **7.** an error   **8.** some multiple-bit errors

SELF-TEST EXERCISE 8–3

1. Exclusive NOR or comparator   **2. a.** 10111   **b.** 10010
c. 10000110   **d.** 1111   **e.** 111111   **f.** 10.01   **3.** carry out
4. carry in   **5.** 7 full adders and 1 half adder   **6. b.** all bits are
added sequentially, LSB first

SELF-TEST EXERCISE 9–1

1. b. an analog-to-digital converter   **2. a.** a digital-to-analog
converter   **3.** high-gain dc amplifier

4.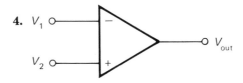

5. high   **6.** low

SELF-TEST EXERCISE 9–2

1. increased   **2.** stability   **3.** $-(10\,\text{k}\Omega/1\,\text{k}\Omega) = -10$ **4.** $-10\,\text{V}$
5. $(20\,\text{k}\Omega/2\,\text{k}\Omega) + 1 = 11$ **6.** $16\,\text{V}/11 = 1.45\,\text{V}$

SELF-TEST EXERCISE 9–3

1. $V_{\text{out}} = -R_{\text{f}}[V_1/R_1) + (V_2/R_2) + (V_3/R_3)]$
$\quad\quad = -1\,\text{k}\Omega[(5\,\text{V}/1\,\text{k}\Omega) + (2\,\text{V}/2\,\text{k}\Omega) + (3\,\text{V}/4\,\text{k}\Omega)]$
$\quad\quad = -1\,\text{k}\Omega\,(0.005 + 0.001 + 0.00075)$
$\quad\quad = -6.75\,\text{V}$
2. the full gain   **3.** $-5\,\text{V}$ **4.** $+5\,\text{V}$

SELF-TEST EXERCISE 9–4

1. True   **2.** $128 \times 0.05 = 6.4\,\text{V}$ **3.** $2 - 1 \times \text{LSB} = 12.75\,\text{V}$
4. $V_{\text{out}} = -V_{\text{ref}}\,(R_{\text{f}}/R_{\text{b}} + R_{\text{f}}/R_{\text{c}})$
$\quad\quad = -10\,\text{V}[(8\,\text{k}\Omega/20\,\text{k}\Omega) + (8\,\text{k}\Omega/40\,\text{k}\Omega)]$
$\quad\quad = -10\,\text{V}[(0.4) + (0.2)]$
$\quad\quad = -10\,\text{V}(0.6)$
$\quad\quad = -6\,\text{V}$
5. $V_{\text{out}} = -[(V_{\text{ref}}/2^N)\,(\text{decimal value of binary input})\,(R_{\text{f}}/R)]$
$\quad\quad = -[+5\,\text{V}/2^4)\,(1100_2)\,(10\,\text{k}\Omega/50\,\text{k}\Omega)]$
$\quad\quad = -[(0.3125)\,(12)\,(0.2)]$
$\quad\quad = -0.75\,\text{V}$
6. $I_{\text{out}} = I_{\text{ref}}[(1/2^N)\,(\text{decimal value of binary input})]$
$\quad\quad = 2\,\text{mA}\,(1/2^4)\,(0100_2)$
$\quad\quad = 2\,\text{mA}\,(1/16)\,(4)$
$\quad\quad = 500\,\text{A or } 0.5\,\text{mA}$
7. $I_{\text{out}}\,\text{max} = 2\,\text{mA}\,[(1/2^N)\,(1111_2)] = 2\,\text{mA}\,(1/16)\,(15) = 1.875\,\text{mA}$

SELF-TEST EXERCISE 9–5

1. analog   **2.** True   **3.** $\text{LSB} = V_{\text{ref}}\,(2^N) = 10\,\text{V}/2^{10} = 10\,\text{V}/1024 =$
$9.8\,\text{mV}$ **4.** LSB resolution $= 15\,\text{V}/2^8 = 0.0586\,\text{mV}$; digital output $=$
$V_{\text{a}}/\text{LSB resolution} = 6.8\,\text{V}/0.0586 = 116.04$; binary output $= 01110101_2$
5. conversion time $= 93 \times (1/F) = 93 \times (0.5\,\mu\text{s}) = 46.5\,\mu\text{s}$ **6.** LSB
resolution $= 10\,\text{V}/2^4 = 10/16 = 0.625\,\text{V}$; digital output $= 6.8\,\text{V}/0.625 =$
10.88; binary output $= 1011$ **7.** conversion time $= N \times (1/F) =$
$4 \times 0.5\,\mu\text{s} = 2\,\mu\text{s}$

SELF-TEST EXERCISE 10–1
1. Access time   **2.** loses its data   **3.** write   **4.** read
5. random-access memory   **6.** Dynamic   **7.** address

SELF-TEST EXERCISE 10–2
1. volatile   **2. a.** battery backup   **3.** nonvolatile   **4.** False
5. random-access   **6.** 4096   **7.** 65,536 (8192 × 8)   **8.** 1024
9. four

SELF-TEST EXERCISE 10–3
1. random-access   **2.** not erasable   **3.** PROMs   **4.** ultraviolet
light   **5.** electrically alterable read-only memory   **6.** nonvolatile
7. read/write   **8.** True   **9.** ROMs

SELF-TEST EXERCISE 10–4
1. 64 (2^6)   **2.** 1024 (256 × 4)   **3.** False   **4.** one
5. selector

SELF-TEST EXERCISE 10–5
1. power   **2.** address   **3.** False   **4.** low, high, and high impedance (0, 1, Z)   **5.** bus   **6.** cannot

Index